Beyond Pluto
Exploring the outer limits of the solar system

In the last ten years, the known solar system has more than doubled in size. For the first time in almost two centuries an entirely new population of planetary objects has been found. This 'Kuiper Belt' of minor planets beyond Neptune has revolutionised our understanding of how the solar system was formed and has finally explained the origin of the enigmatic outer planet Pluto. This is the fascinating story of how theoretical physicists decided that there must be a population of unknown bodies beyond Neptune and how a small band of astronomers set out to find them. What they discovered was a family of ancient planetesimals whose orbits and physical properties were far more complicated than anyone expected. We follow the story of this discovery, and see how astronomers, theoretical physicists and one incredibly dedicated amateur observer have come together to explore the frozen boundary of the solar system.

JOHN DAVIES is an astronomer at the Astronomy Technical Centre in Edinburgh. His research focuses on small solar system objects. In 1983 he discovered six comets with the Infrared Astronomy Satellite (IRAS) and since then he has studied numerous comets and asteroids with ground- and space-based telescopes. Dr Davies has written over 70 scientific papers, four astronomy books and numerous articles in popular science magazines such as *New Scientist*, *Astronomy* and *Sky & Telescope*. Minor Planet 9064 is named Johndavies in recognition of his contributions to solar system research.

Beyond Pluto

Exploring the outer limits
of the solar system

John Davies

CAMBRIDGE
UNIVERSITY PRESS

CAMBRIDGE UNIVERSITY PRESS
Cambridge, New York, Melbourne, Madrid, Cape Town,
Singapore, São Paulo, Delhi, Tokyo, Mexico City

Cambridge University Press
The Edinburgh Building, Cambridge CB2 8RU, UK

Published in the United States of America by Cambridge University Press, New York

www.cambridge.org
Information on this title: www.cambridge.org/9781107402614

First published 2001
First paperback edition 2011

A catalogue record for this publication is available from the British Library

Library of Congress Cataloguing in Publication data
Davies, John Keith.
Beyond Pluto : exploring the outer limits of the solar system / John Keith Davies.
  p. cm.
Includes bibliographical references and index.
ISBN 0 521 80019 6
1. Kuiper Belt. I. Title.
QB695.D38 2001
523.2–dc21 00-049364

ISBN 978-0-521-80019-8 Hardback
ISBN 978-1-107-40261-4 Paperback

To all my friends in Hawaii, especially those who made UKIRT the greatest infrared telescope in the world.

Contents

Preface

This is a story about a discovery and some of the developments which followed it. It is not a textbook. Although I hope it contains most of the relevant technical details I set out to show a little of how astronomy is actually done. Some of the characters spend their time looking through telescopes on the darkest of dark nights, others work in offices and laboratories far removed, both physically and psychologically, from mountaintop observatories. From time to time this diverse group of people come together, in small groups or *en masse*, to exchange ideas and dispute data. They do this in order to understand the origin and evolution of the solar system in which we live and work. A few names crop up frequently, for the community of solar system astronomers is a small one and our paths often meander across each other in unpredictable ways.

In the last few years a new, and dynamic, outer solar system has replaced the sterile border known to our predecessors. I still find it hard to believe how much our view of the solar system has changed in the last decade and even harder to credit that I have been a part of this adventure. It has been an exciting time for all of us, and some of my childhood dreams have come true in a way that I could never have imagined. I hope that some of this mystery and excitement comes through in these pages.

Acknowledgements

Although I put this book together, many of the ideas in it sprang from the fertile minds of my fellow astronomers. It is they who have opened up the distant frontier of the solar system. Many of them have shared their thoughts, and in some cases their images, with me over the last couple of years. Some consented to be interviewed in person, others tolerated a barrage of email enquiries, and all of them seemed to have taken it in good humour. A good fraction of them read specific sections of the draft manuscript (a few brave souls tackled the whole thing) and put me right when I strayed from the facts. They also helped make clear to me things that were uncertain. I enjoyed the spirited debate about the planetary status of Pluto, and the sharing of recollections or thoughts about events that happened a long time ago. I'd like to thank them all for their help. In particular:

Mike A' Hearn, Mark Bailey, Gary Bernstein, Alan Boss, Mike Brown, Robert (Bob) Brown, Marc Buie, Al Cameron, Anita Cochran, Dale Cruikshank, Cathy Delahodde, Martin Duncan, Julio Fernandez, Tom Gehrels, Brett Gladman, Dan Green, Simon Green, Jane Greaves, Oli Hainaut, Eleanor (Glo) Helin, James Hilton, Wayne Holland, Piet Hut, Dave Jewitt, Charles Kowal, Jane Luu, Renu Malhotra, Brian Marsden, Neil McBride, John McFarland, Steve Miller, Warren Offutt, Joel Parker, Dave Rabinowitz, Eileen Ryan, Glenn Schnieder, Jim Scotti, Carolyn Shoemaker, Ronald Steahl, Alan Stern, Steve Tegler, Rich Terrile, Dave Tholen, Nick Thomas, Chad Trujillo, Tony Tyson and David Woodworth.

Material in Appendix 1 is reproduced with the permission of the Minor Planet Center and Lutz D. Schmadel. A full list of Minor Planet citations may be found in the *Dictionary of Minor Planet Names*, by L. D. Schmadel, which is published by Springer Verlag and is currently in its 4th edition.

Many of the illustrations came directly from the scientists involved and are credited in the captions. Figures 3.8, 6.2, 7.2, 7.4 and 8.1 are reproduced from *Protostars and Planets* IV (editors Vince Mannings, Alan Boss, Sara Russell) with permission of the University of Arizona Press, ©2000 The Arizona Board of Regents, courtesy of Dave Jewitt. Figure 2.4 was published in *Icarus* and appears with permission of Academic Press. I would also like to thank the following for their help with specific images or information.

George Beekman, *Zenit Magazine.*

Germaine Clark & Brooke Hitchcock, Lawrence Livermore National Laboratory.

Richard Dreiser, Yerkes Observatory.

Cheryl Grundy, Space Telescope Science Institute.

Debora Hedges & Lynne Nakauchi, California Institute of Technology.

Bob Milkey, American Astronomical Society.

Lori Stiles, University of Arizona.

Richard Wainscoat, Institute for Astronomy, Honolulu.

Cynthia Webster, Lowell Observatory.

Remo Tilanus for translating George Beekman's article into a language I understand and to Tim Jenness for converting those GIFs, TIFFs, JPEGs and bitmaps into things my printer could understand.

Adam Black, of Cambridge University Press for getting me into this, and to Alice Houston, Simon Mitton and Alison Litherland, also of the Press, who got me out of it.

Finally, my wife Maggie for reading and correcting the manuscript, even though it's not really her subject.

Prologue

In July 1943 the *Journal of the British Astronomical Association* published a short article entitled 'The Evolution of our Planetary System'. The paper had been submitted by a retired Irish soldier and part-time amateur theoretical astronomer, Lt-Col. Kenneth Edgeworth. Despite being greatly reduced in length due to wartime shortages of paper, the article contained a prophetic paragraph on the structure of the solar system. While discussing comets, Lt-Col. Edgeworth remarked, 'It may be inferred that the outer region of the solar system, beyond the orbits of the planets, is occupied by a very large number of comparatively small bodies.' Kenneth Edgeworth did not live to see his prediction confirmed, but almost 50 years later just such an object was discovered. This new body, initially called simply 1992 QB$_1$, was the harbinger of a breakthrough in our understanding of the solar system. Within a few years hundreds of similar objects would be found in what, by an ironic twist, soon became known as the Kuiper, rather than Edgeworth, Belt.

The edge of the solar system

Like the planets Pluto and Neptune, the existence of what is frequently called the Kuiper Belt was predicted theoretically long before it was actually observed. Probably the first fairly detailed speculation about a cometary ring beyond Neptune was put forward by Kenneth Essex Edgeworth in 1943. Edgeworth was an interesting character who had progressed from soldier and engineer to retired gentleman and amateur theoretical astronomer. He was born on 26th February 1880 in County Westmeath, Ireland, into a classic well to do literary and scientific family of that era. As a young man he joined the Royal Military Academy at Woolwich, England, and attained a commission in the Royal Engineers. He spent his next few years stationed around the world building bridges, barrack blocks and the like. With the outbreak of the First World War he served with the British Army in France as a communications specialist and was decorated with both the Distinguished Service Order and the Military Cross. He remained in the army until 1926 and then took up a position with the Sudanese department of Posts and Telegraphs in Khartoum. Edgeworth remained in the Sudan for five years before retiring to Ireland to live out the remainder of his life.

Although retired, Edgeworth was by no means inactive. During the 1930s he studied economic theory and published several books on this topic. Although never affiliated with a university or other astronomical institution, he also pursued an interest in astronomy which he had acquired in his youth (he had joined the Royal Astronomical Society in 1903). After he retired he wrote a number of articles, mostly theoretical in nature, dealing with the process of star formation and developing ideas about the origin of the solar system. In 1943 he joined the British Astronomical Association, whose journal published his first paper on the evolution of the solar system that summer. It was a short note which Edgeworth himself described as containing 'Not so much a theory, but the outline of a theory with many gaps remaining to be filled'. In his paper he described how a cloud of interstellar gas and dust might collapse to form a disc. He speculated that within such a disc numerous local condensations of higher density might then develop and collapse upon themselves. Noting that the real solar system does not comprise a huge number of small objects, but rather a few large planets and moons, Edgeworth suggested that these condensations then coalesced to form the nine known planets and their satellites. Crucially, Edgeworth recognised that there was no obvious reason why the disc of planet-forming material should have been sharply bounded at the orbit of the outermost planet. He suspected that the disc probably extended far beyond this distance and reasoned that, so far from the Sun, the density of material in the disc would be

Figure 1.1 A caricature of Kenneth Edgeworth as a young man. Comparison with photographs of him in later life suggests that it is a good likeness. (Royal Signals Museum Archive.)

very low. So, although individual condensations of reasonable size might form beyond Neptune, there would be little likelihood of them encountering each other frequently enough to form large planets. He suggested instead that these condensations would simply collapse upon themselves to form a large number of small bodies. Echoing then current theories of comets as concentrated swarms of meteoroids he described these distant condensations as astronomical heaps of gravel. He added that perhaps from time to time one of these condensations 'Wanders from its own sphere and appears as an occasional visitor to the inner solar system'. Here was the genesis of the idea of a trans-Neptunian comet belt.

Edgeworth developed his ideas further, writing a longer paper along similar lines a few years later. This second paper was submitted to the *Monthly Notices of the Royal Astronomical Society* in June 1949. Although by today's standards it contained numerous poorly justified assumptions, it was accepted almost immediately and appeared in an issue of the journal dated late 1949. In this paper Edgeworth expanded on his model for the formation of the planets and once again mentioned the likely existence of a vast reservoir of potential comets beyond the orbit of Neptune.

About the same time as Edgeworth's musings, the Dutch-born astronomer Gerard Kuiper was also considering the existence of tiny worlds beyond Pluto. Kuiper was working at the Yerkes Observatory in Chicago and, in 1951, he wrote what became a classic book chapter summarising the state of knowledge about the solar system. Kuiper noted that the distribution of material in the outer solar system seemed to come to an unnaturally sharp edge in the region of the planet Neptune and that there was no obvious reason why this should be so. Perhaps taking a lead from newly published theories about the composition of comets, Kuiper suggested that during the formation of the planets many thousands of kilometre-sized 'snowballs' might have been formed in a disc beyond the planet Neptune. Like Edgeworth, Kuiper reasoned that at such great distances from the Sun, where the relatively tiny snowballs would occupy a huge volume of space, it was unlikely that these snowballs could come together to form large planets. He suggested that instead their orbits were disturbed by the gravitational influence of the planet Pluto[†] and they were either ejected into deep space or sent in towards the Sun to

---

[†] Pluto was then thought to be much more massive than we now know it to be.

appear as comets. However, in a world in which observational astronomy was still dominated by the photographic plate, the detection of such tiny objects remained impracticable.

Of course, speculation about missing planets is not a new phenomenon. Ever since William Herschel's discovery of Uranus in 1781, astronomers have been fascinated by the possibility that there might be other unknown worlds. On the 1st of January 1801 the Italian astronomer Giuseppe Piazzi made a chance discovery of what was at first thought to be a new planet. The object, which was soon shown to be orbiting between Mars and Jupiter, was named Ceres after the Roman goddess of the harvest. It was soon found that Ceres, even though it was quite close, did not show a detectable disc when viewed through a telescope. This suggested that it was smaller than any of the other known planets. Three similar objects, Pallas, Juno and Vesta, were found in 1802, 1804 and 1807 respectively. All appearing as slow-moving points of light, this group of new objects was referred to as asteroids (star-like) by William Herschel. All went quiet for a while until the mid 1840s when new asteroids began to be found in quite large numbers. By the end of 1851 fifteen of them had been found and we now know that Ceres is just the largest of many small rocky objects in what became known as the asteroid belt.

However, by the middle of the nineteenth century attention had once again swung to the outer solar system. Irregularities in the motion of Uranus hinted that it was being tugged by the gravitational pull of another, more distant world still waiting to be discovered. In a now classic story of astronomical detective work, the mathematicians Urbain Le Verrier of France and John Couch Adams of England independently calculated the position of the unseen planet, making its discovery a relatively simple matter once someone could be persuaded to look in the appropriate direction. In the event, it was Le Verrier whose prediction was first tested. While Adams' calculations lay almost ignored by the English Astronomer Royal, the director of the Berlin Observatory J. G. Galle and his assistants searched the region suggested by Le Verrier. They found the predicted planet on 23rd September 1846. However, the discovery of Neptune was not the end of the issue as far as distant planets were concerned. After a few decades it seemed that Neptune alone could not explain all the problems with the orbit of Uranus. This hinted that there might be yet another planet lurking in the darkness of the outer solar system. Two Americans set out to see if this was the case.

William Pickering was one of these planet hunters, suggesting in 1908 that a planet with twice the mass of the Earth should lie in the direction of the constellation Cancer. His prediction was ignored. Eleven years later he revised his calculations and pointed to a position in nearby Gemini. This time astronomers at the observatory on Mt Wilson, California, responded, using their 24 cm telescope to search around Pickering's predicted coordinates. They failed to find anything. Meanwhile, American millionaire Percival Lowell was also turning his attention to the outer solar system. Lowell, who had earlier convinced himself that intelligent life existed on the planet Mars, firmly believed that deviations from the predicted positions of Uranus meant that there must be another unseen planet remaining to be discovered. He called this distant object 'Planet X' and, like Pickering, he tried to calculate where in the sky it might be found. However, Lowell had an advantage over his rival, for he had the means to pursue his search without relying on the whims of others. Lowell owned a private observatory which he had founded in 1894. It was built on Mars Hill, just outside the town of Flagstaff, Arizona. Unlike modern observatories, which are usually located on barren mountaintops, Lowell placed his telescopes in a delightful setting. The Lowell observatory was surrounded by pine trees and had a fine view back across the town.

Lowell's Planet X was predicted to be quite large, but very distant, and so was unlikely to show an obvious disc in the eyepiece of a small telescope. The best way to find it would be to detect its daily motion relative to the fixed background of stars and galaxies. In the previous century such searches had been made by laboriously sketching the view through a telescope and then comparing this with sketches of the same region made a few days earlier. However, by Lowell's time, astronomical photography had come on the scene and offered a much faster and more reliable way to survey the sky. Lowell's first search was made between 1905 and 1907 using pairs of photographic plates which he scanned by eye, placing one above the other and examining them with a magnifying glass. He soon realised that this method was not going to work.

Lowell's next step was to order a device known as a blink comparator to assist in the examination of the photographs. The comparator provided a magnified view of a portion of the photographs but, more importantly, it allowed the searcher to switch rapidly between two different images of the same patch of sky. Once the photographs were aligned correctly, star images remained stationary as the view flashed from one plate to the other. However, should there be a moving object in the field of view, its image would jump backwards and forwards as the images were interchanged. Naturally enough, the process was known as 'blinking' the plates.

A search of the constellation Libra was made in 1911, but was abandoned after a year when nothing was found. Undeterred by this failure, Lowell began another search in 1914. Between then and 1916 over 1000 photographic plates were taken, but once again nothing was found.[†] Lowell died suddenly from a stroke on the 16th of November 1916, his planet-finding ambition unfulfilled. He was buried in a small mausoleum, shaped to resemble the planet Saturn, in the grounds of his observatory on Mars Hill. For a time the search for Planet X was halted as Lowell's widow tried to break the provisions of his will. Mrs Lowell wanted to remove funds from the operation of the observatory and preserve the site as a museum in her late husband's memory. The resulting litigation siphoned off funds from the observatory for a decade.

Eventually, under the directorship of Vesto Slipher, the Lowell Observatory returned to the problem of the missing planet. Slipher recruited a young amateur astronomer named Clyde Tombaugh, a farmboy from Kansas, as an observing assistant. Tombaugh arrived in Flagstaff during January 1929 and was set the task of taking photographs which could be searched for Lowell's Planet X. It took a while to get the new 31 cm telescope, built especially for the search, into full operation, but by April all was ready. Tombaugh took a number of photographic plates covering the region around the constellation of Gemini, the latest predicted location of Planet X. The plates were 33.5 cm by 40 cm in size and covered nearly 150 square degrees of sky. Each contained many thousands of star images. Vesto Slipher and his brother blinked the plates over the next couple of weeks, but they failed to find anything. In the meantime, Tombaugh continued to photograph the sky and soon a large backlog of unexamined plates had built up. Slipher then asked Tombaugh to blink the plates as well as taking them, explaining that the more senior observatory staff were too busy to devote much time to the onerous and time-consuming blinking process.

[†] In fact the missing planet did appear on two of these images, but it was much fainter than expected and its presence was missed.

Figure 1.2 Clyde Tombaugh entering the dome of the Lowell Observatory's 33 cm telescope. He is carrying a holder for one of the photographic plates. After exposing the plate he had to search it millimetre by millimetre for Planet X. Few astronomers now go to their telescopes so formally dressed, as can be seen by comparison with figures 4.1 and 5.7. (Lowell Observatory archives.)

Tombaugh regarded the prospects of his new assignment as 'grim', but he dutifully continued with his programme. Night after night he made a systematic photographic survey of the sky. He concentrated on regions close to the ecliptic, an imaginary line across the sky which marks the path traced out by the Sun across the constellations of the zodiac during the course of a year. The ecliptic is not the precise plane of the solar system, which is better defined by taking account of all the planets and not just of the Earth. When this is done the result is known as the invariable plane. However, when projected onto the sky, the ecliptic and the invariable plane are not much different and it is common, if careless, to use the two terms interchangeably. Since the orbit of Planet X would presumably be close to the invariable plane, the ecliptic was the obvious region around which to search.

Tombaugh's method was to take three photographs of each region of sky at intervals of two or three days. Each photograph was exposed for several hours. During each exposure Tombaugh painstakingly guided the telescope to make sure that the images of the stars were sharp, with all their light concentrated onto as small an area of the photographic emulsion as possible. Only then would his plate reveal the very faintest objects and give him the best chance of success. At dawn he developed the plates, careful lest a tiny mistake ruin them and waste his hours of work in the telescope dome. Later he examined the plates for anything which might have moved between the two exposures.

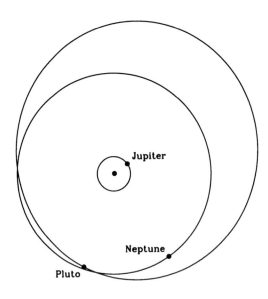

Figure 1.3 The orbits of Jupiter, Neptune and Pluto. Pluto's eccentric orbit crosses that of Neptune, although the significance of this was not realised at the time of its discovery. (Chad Trujillo.)

Although the technique sounds simple in principle, Tombaugh's task was a huge one. The long nights of observing were tiring and the blinking of the frames was tedious in the extreme. Many false detections appeared, caused by things such as variable stars, chance alignments of main belt asteroids and photographic defects which mimicked moving objects. To eliminate these false detections, Tombaugh used his third plate to check if any of the candidate objects were visible again. Usually, of course, they were not. Tombaugh's patience was finally rewarded on the 18th of February 1930 when he was examining a pair of plates he had taken a few weeks earlier. Blinking them, he found a moving object that was clearly not a star, a nearby asteroid or a flaw in the photographic emulsion. What was more, the object's slow motion across the sky suggested that it must be well beyond Neptune. After a few more weeks of observations had been made to define the object's motion more accurately, the discovery was announced on 13th March. The date was chosen since it would have been Lowell's 75th birthday if only he had lived to see the day. After a certain amount of debate, to which we shall return later, the new object was named Pluto, after the god of the underworld.

Clyde Tombaugh continued his search for another 13 years. He estimates that in this time he covered about 70% of the heavens and blinked plates covering some 90000 square degrees of sky.[†] All in all he spent some 7000 hours scanning every square millimetre of about 75 square metres of plate surface. Although he marked 3969 asteroids, 1807 variable stars and discovered a comet, he never found another object as distant as Pluto. This was a little odd since it gradually became clear that the new planet was rather smaller than predicted. The first clue that Pluto was small came from its faintness, which suggested it could not be any larger than the Earth. Worse still, even the largest telescopes of the day could not resolve Pluto and show it as a disc. Under even the highest magnifications, the planet remained a tiny point of light, devoid of any features. This was worrying since if Pluto was very small it could not affect the orbit of Uranus to any significant extent. None the less, the intensity of Tombaugh's efforts seemed to rule out any chance that any other massive planet could exist near Neptune's orbit.

It was not until much later that theoretical work, notably by

[†] The total area of the sky is less than this, but some regions were examined more than once.

American E. Myles Standish in 1993, explained the apparent deviations in the motion of Uranus. Standish based his calculations on improved estimates of the masses of the giant planets which had been determined during the flybys of the Voyager spacecraft. Using these he showed that any remaining errors in the measurements of Uranus' position were tiny and could be explained by systematic observational uncertainties. There was no need to invoke the gravitational influence of a missing planet, massive or otherwise. Lowell's hypothesis of a Planet X had been completely wrong. The discovery of Pluto was a consequence of the thoroughness of Tombaugh's systematic search and the fact that Pluto was fairly close to Lowell's predicted position was just a coincidence.

It was well into the 1970s before the true nature of Pluto was revealed. The planet's orbit was quite well defined within a year of its discovery, but Pluto's faintness made determining details of its physical make-up almost impossible for decades. In the mid 1950s it was established that Pluto has a rotation period of 6.39 days and in 1976 methane frost was detected on its surface. Since methane frost is quite reflective, this implied that Pluto was even smaller than at first thought. Pluto soon shrank again. In 1977 James Christy was examining images of Pluto when he noticed that the planet seemed to be elongated some of the time and not others. He soon realised that this was due to the presence of a large satellite going around the planet every 6.39 days, the same as Pluto's rotation period. As its discoverer, Christy had to name the new moon and he chose Charon, the name of the ferryman who transported souls to the underworld. Strictly speaking Charon should be pronounced Kharon, but it is often enunciated as Sharon since Christy's wife, Sharlene, is known to her friends as Shar. Once the details of Charon's orbit had been established, it was possible to determine the combined mass of Pluto and Charon. This turned out to be no more than 0.0024 times the mass of the Earth. Pluto was a small and icy world. Although the true size of Pluto was unclear in the 1940s, it may have been the realisation that there was no massive Planet X that made Edgeworth and Kuiper speculate about the edge of the solar system. Certainly the existence of small icy worlds at the fringe of the planetary region seemed a natural conclusion from theories of how the solar system formed.

It had once been suggested that the solar system was produced when a close encounter between our Sun and another star pulled out a filament of material which condensed into planets. However, it was soon shown that this could not be the case. The realisation that the distances between the stars were very large made such an encounter unlikely, but more importantly, it can be shown mathematically that material pulled out from the Sun could not form planets. Ejected material would either fall back onto the Sun or disperse into space. So astronomers rejected this near-encounter model. Instead, they embraced an idea put forward by the philosopher Immanuel Kant in 1755 and subsequently developed by a French scientist, Pierre Simon, Marquis de Laplace. In 1796 Laplace suggested that the Sun formed in a slowly rotating cloud of gas and that, as the cloud contracted, it threw off rings of material which formed the planets. Although many of the details have been improved, the general outline of this nebular hypothesis survives today.

Modern theories of the formation of our solar system begin from the assumption that stars like the Sun form in the clouds of gas and dust which exist throughout interstellar space. These clouds often contain as much as a million times more mass than the Sun and each spreads over a huge volume of space. From time to time, instabilities develop within these clouds and bursts of star formation are triggered. About five billion years ago, an instability in just such a cloud triggered one such collapse. At the centre of this collapsing region, itself buried deep within the larger interstellar cloud, a dense clump of material began to form. As this protostellar core contracted, it increased in mass and so generated a more powerful gravitational field. This in turn attracted in more material, increasing the mass of the core still further in a rapidly accelerating process. As material fell in towards the centre it was slowed down by friction and gave up its kinetic energy as heat, gently warming the central regions of the core. For a while, the heat could leak out in the form of infrared and sub-millimetre radiation and so the collapsing core remained quite cool. However, as the cloud got more and more dense, a point was reached when its central regions became opaque to most forms of radiation. When this happened, heat could no longer escape easily and the temperature at the centre began to rise rapidly. After a while conditions reached the point at which nuclear reactions could begin and the core began to convert hydrogen to helium. The energy released by these nuclear reactions generated sufficient pressure to halt any further collapse and the star we call the Sun was born.

Of course the details of the star formation process are far more complicated than can be described in a single paragraph. In particular, a mathematical analysis of the fate of a spherical collapsing cloud immediately throws out a simple, but vitally important question. If the Sun formed from the collapse of a huge cloud of gas, why does it rotate so slowly, taking about 11 days to turn on its axis? This fact alone hints at the existence of planets as a consequence of the physical law that angular momentum, or spin energy, must be conserved.

The conservation of angular momentum can be observed when an ice dancer skating with arms outstretched enters a tight turn and begins to spin on the spot. If, as her spin begins to slow down, the dancer brings her arms in close to her body, her rate of rotation suddenly speeds up. A similar effect can be experienced, without getting cold feet, by sitting on a well oiled office chair and spinning around on it with your arms held out. If you pull in your arms you can feel the spin rate increase, push them out again and the spinning slows down. Try again with a heavy book in each a hand and you will find it works even better. This simple observation is revealing two important things about physics. Firstly, angular momentum depends on both the rate at which something is spinning and upon the distance of its constituent masses from the axis of rotation. Secondly, the total amount of angular momentum in a spinning system is conserved. So, as demonstrated by our ice dancer, as mass is brought in towards the axis of rotation of a spinning system, the spin rate must increase to keep the total amount of angular momentum, or spin energy, the same. The more mass there is on the outside of the spinning region, and the further the mass is from the spin axis, the more angular momentum the system has.

The problem faced by the forming Sun was that as the protostellar cloud collapsed, it had to lose considerable amounts of angular momentum. This is necessary because unless the original cloud was completely at rest when the infall began, then as material fell inwards, it would have transferred its angular momentum to the central regions. This would have increased the rate of rotation of the proto-sun quite dramatically. Unless this angular momentum could be removed, the spin rate would continue to increase as the collapse proceeded. By the time the core had shrunk to stellar dimensions, the rotation would be far too rapid to allow a star to form. So, somehow during its collapse, the core must have transferred angular momentum to material further out in the cloud. This occurred as magnetic fields and gas drag gradually forced the outer reaches of the cloud to spin around with the core. As this continued, the outer regions of what had been a spherical cloud fell in towards the equator and the nebula became a huge, slowly spinning disc surrounding a small stellar embryo. The forming Sun continued to grow as material in the disc fell inwards onto it.

The conditions across the protoplanetary disc depended on the balance between the energy generated during the collapse, the light emitted by the still-forming Sun and the rate at which energy was transported through the disc. In the central regions it was too hot for icy material to survive. Here, in what became the inner solar system, most of the ices were evaporated and blown outwards, leaving behind more robust dusty material. Within the spinning disc, tiny grains began to bump into each other. The grains, remnants of the original interstellar cloud, were probably smaller than a micron across to start with, but the collisions were gentle enough that they began to stick together. At first they formed fluffy structures which were mostly empty space, but as they grew still further, they began to compact. Soon they reached the point were they were more like small pebbles, jostling each other as they orbited the Sun. Inexorably these lumps of debris grew still further. Then, once a few objects had reached a size of about ten kilometres in diameter, a dramatic change of pace occurred.

These larger lumps, or planetesimals, were now massive enough that their gravitational fields began to attract other passing material onto themselves. Once this started it dramatically accelerated the growth process. Before long a few planetesimals began to dominate all of the space around them, clearing away the remainder of the orbiting material by dragging it down onto their surfaces. Within 100 000 years or so many rocky bodies about the size of the Earth's moon had formed. After this brief period of runaway growth, the pace slowed again. By now each planetary embryo had swept up all the material within reach and the distances between the larger objects were too great for them to encounter each other. It took another 100 million years for the planet-building process to be completed. Gradually, subtle gravitational interactions between the planetesimals stirred up their orbits enough for occasional dramatic collisions to occur. One by one the surviving embryos were swept up into the four terrestrial, or Earth-like, planets, which we see today.

Further out, about half a billion kilometres from the Sun, temperatures were low enough that ices could survive. So, as well as dust, the outer regions of the disc contained considerable amounts of water ice and frozen gases such as methane, ammonia and carbon monoxide.

Here, the growing planetary cores swept up this extra material to form giant planets dominated by the gases hydrogen and helium with a seasoning of various ices. Jupiter, the largest of these giants, was so large that, even while it was still forming, its gravitational field had a dominating effect on its neighbourhood. Jupiter's gravity stirred up the region between itself and the still forming planet Mars and prevented a single object dominating this region. Instead of forming a fully fledged planet, the growth stopped, leaving a population of smaller, rocky asteroids. Beyond Jupiter, the other giant planets Saturn, Uranus and Neptune grew as they too swept up the icy planetesimals from the space around them.

At great distances from the Sun the protoplanetary disc became much more diffuse. Here, although there was sufficient material to reach the stage of forming small planetesimals, there was not enough time, or enough material, for them to combine into a major planet. Instead they formed a diffuse zone of small icy objects in almost permanent exile at the fringes of the solar system. This is the frozen boundary of the planetary region; beyond it lies only the huge, more-or-less spherical cloud of planetesimals ejected into deep space by gravitational interactions with the forming planets and the rest of the stars in our galaxy.

After Edgeworth's and Kuiper's articles, thinking about a possible disc of planetesimals beyond Neptune lapsed until the early 1960s. A brief revival of interest began in 1962 when naturalised American physicist Alistair Cameron[†] wrote a major review about the formation of the solar system. Cameron's review appeared in the first issue of *Icarus*, a new scientific journal devoted exclusively to the study of the solar system. Using the same arguments as Kuiper and Edgeworth, namely that material in the outer regions of the protoplanetary disc would be too diffuse to form a planet, Cameron wrote that 'It is difficult to escape the conclusion that there must be a tremendous mass of small material on the outskirts of the solar system'. Soon after Cameron wrote his review, another astronomer turned his attention to the possible existence of a comet belt beyond Neptune.

American Fred Whipple, who had done much to explain the composition of comets a decade earlier, began by accepting the likely existence of what he called a comet belt. From his knowledge of comets, he reasoned that if a trans-Neptunian belt of icy planetesimals

---

[†] Cameron was originally a Canadian.

existed, then even objects as large as 100 km in diameter would be very faint. This would make the discovery of individual objects highly unlikely with then existing astronomical technology. Turning the question around, he then asked himself if the comet belt would be detectable if the comets within it were very small. Would the combined light of large numbers of small comets produce a faint, but detectable glow across the sky? His conclusion was that any glow from the comet belt would be too faint to see against the background of the night sky. In particular it would be masked by the diffuse glow of the zodiacal light, a band of light along the ecliptic plane produced by sunlight scattering off interplanetary dust in the inner solar system. Having decided that it was impossible to detect a hypothetical comet belt directly, he set out to attack the problem dynamically. Harking back to Adams, Le Verrier and Lowell, Whipple tried to find out if a comet belt could have any measurable gravitational effects on the rest of the solar system.

Whipple first considered the gravitational effect of the belt on the motion of the planets Uranus and Neptune. He concluded that a comet belt having a mass of 10–20 times that of the Earth might exist beyond Neptune, but that the evidence for such a belt was not conclusive. He merely noted that a hypothetical comet belt provided a better explanation of the apparent irregularities in the motion of Neptune than assigning a mass to Pluto that was much larger than seemed justified by other observations of the tiny planet. He even went as far as to say Pluto could not affect the other two planets significantly even if it were made of solid gold. In 1967 Whipple, together with S.E. Hamid and a young astronomer called Brian Marsden, tried to estimate the mass of the comet belt another way. They looked for its effect on the orbits of seven comets which all travelled beyond Uranus. They then calculated the gravitational effect that a hypothetical comet belt containing as much material as the Earth would have on the orbits of each of these comets. Since they found that the real comets had suffered no such effects, they concluded that any unseen comet belt could not have a mass of much more than one Earth mass. Thus Edgeworth's and Kuiper's ideas remained largely in limbo for a number of years. It was only when a number of advances in our understanding of comets began to come together that it was gradually realised that there was a problem that could best be solved by postulating the existence of an ecliptic comet belt.

The existence of comets, as ghostly apparitions that appear without warning, move slowly across the sky and then fade away, has been known throughout history. However, only in the latter half of the twentieth century was a reasonable physical model of a comet developed. Although Edmund Halley noticed the similarity between the orbits of a number of comets, realised they were the same object and predicted the return of what has become known as Halley's Comet, neither he nor his contemporaries really understood what a comet actually was. By the early 1900s the favoured model was of a loose aggregation of dust and rocks, little more than a loosely bound cloud of material, carrying with it gas molecules trapped both on the surfaces of the grains and deep within pores of the larger pieces. When the comet was warmed by the Sun, these gases were apparently released to form a tail. There were serious flaws with this model, the most significant being that such a system could not supply enough gas to explain the rate at which gas was known to be produced as a comet approached the Sun. There was really only one thing that was known for certain about comets: dynamically speaking, they were of two distinct types. Comets of one kind appeared unpredictably from random directions on the sky and made a single trip around the Sun before disappearing for thousands of years. Those of the other kind, which were generally much fainter, reappeared regularly and their returns could be predicted quite accurately.

Comets of the first kind, called long-period comets, follow very elongated (parabolic) orbits which range from the inner solar system at one extreme into deep space at the other. As with most solar system objects, it is convenient to describe these orbits in terms of astronomical units (AU). An astronomical unit is defined as the average distance of the Earth from the Sun and amounts to about 150 000 000 km. Using these units, Jupiter is 5 AU from the Sun and Neptune's orbit is at about 39 AU. The long-period comets which can be observed from Earth have perihelia, or closest approaches to the Sun, of less than a few astronomical units and aphelia, or furthest distances from the Sun, of many thousands of AU. In the late 1940s the Dutch astronomer Jan Oort examined the statistics of the few hundred long-period comets then known and suggested that they came from a huge, more-or-less spherical shell around the Sun which extended to about half way to the nearest stars. Oort believed that the comets were ancient planetesimals that had been gravitationally ejected from the region of what is now the asteroid belt during the planet-building process about

16

four and a half billion years ago. He suggested that they then remain in the distant cloud until random gravitational forces from other nearby stars change their orbits slightly and cause them to fall inwards towards the warm heart of the solar system. Gerard Kuiper, an ex-student of Oort's, soon pointed out that the comets, being icy, were probably formed in the region from 35 to 50 AU rather than in the asteroid belt. Kuiper believed they were ejected by Pluto, not Jupiter. However, the broad outline of Oort's theory for the origin of comets was generally accepted and the hypothetical shell of distant comets became known as the Oort Cloud.

About the same time as Oort was explaining the dynamics of the long-period comets, Fred Whipple brought forward his icy conglomerate or 'dirty snowball' model of a comet. He suggested that the essence of a comet was a single solid body a few kilometres across called the nucleus. Each comet nucleus comprises frozen ices such as water, carbon monoxide, ammonia and methane together with a small amount of dust. As the nucleus approaches the Sun, solar heating warms it and causes the frozen gases in its outer layers to sublime. This creates a physically large, but very tenuous cloud around the nucleus. This cloud is called the coma. The coma is not entirely gas, since as the gases leave the nucleus they carry with them tiny dust particles. The pressure of sunlight, and the solar wind of material

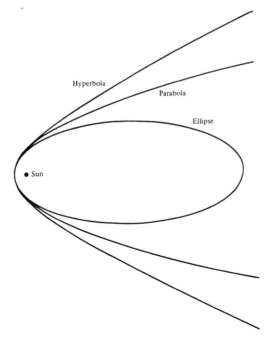

Hyperbola

Parabola

Ellipse

Sun

Figure 1.4 Comet orbits. Long-period comets approach the Sun along parabolas. Short-period comets orbit the Sun in ellipses usually, but not always, quite close to the plane of the planets. A comet on a hyperbolic orbit would be approaching from outside the solar system. The fact that no such hyperbolic comets are seen is evidence that comets are part of the Sun's family. The elliptical orbit has an eccentricity of 0.9.

constantly flowing out from the Sun, act on the coma and blow material away to form the comet's tail. Most comets actually have two tails, a long straight bluish one comprising gases that have been ionised and are moving directly away from the Sun and a curved, yellowish one comprising individual dust grains being blown away from the nucleus into independent solar orbits. In most comets, depending on the ratio of gas to dust in the nucleus, one of these tails is much more prominent than the other. Sunlight reflected from the coma and the dust tail makes the comet visible from the Earth. Once the comet has passed around the Sun and begins to recede back into deep space, the nucleus cools and the sublimation of the ices slows down and finally stops. Once this happens the comet rapidly becomes too faint to detect. Unseen, the frozen nucleus returns to the Oort Cloud from where, thousands of years hence, it may return to visit the Sun again.

The second class of comets are those of short period. These are confined to the inner solar system and most of them travel around the

Figure 1.5 Comet Hale–Bopp seen from Mauna Kea, Hawaii. Comet Hale–Bopp is a long-period comet and displayed a long, bright dust tail. Short-period comets are almost never visible to the unaided eye. (John Davies.)

Sun in elliptical orbits with periods of about a dozen years. In general, they have orbits of low inclination, that is to say they are close to the plane of the solar system. The short-period comets are generally much fainter than comets coming from the Oort Cloud. This is because short-period comets approach the Sun very frequently and on each trip more and more of the volatile ices which form the coma and tail are removed. So even when heated by the Sun at perihelion, short-period comets are pale shadows of their fresh, bright cousins making their rare appearances in the inner solar system. The faintness of the short-period comets indicates that they are gradually running out of volatile material and that they cannot survive for long in their present orbits. The short-period comets are fated to fade away completely, and to do so quite quickly in astronomical terms. By estimating how much material is removed on every trip around the Sun, astronomers have shown that short-period comets cannot survive in their present locations for even a small fraction of the age of the solar system. However, the very fact that numerous short-period comets do exist means that new ones must be arriving regularly to top up the present supply and replace them as they vanish. Many of these short-period comets have aphelia in the region of Jupiter's orbit; these are called Jupiter family comets. This link with the giant planet is a clue to their origin. Comets from the Oort Cloud which happen to approach Jupiter too closely have a chance of being captured into the inner solar system by Jupiter's gravity, making them doomed to make frequent approaches to the Sun until they disappear forever.

Although the Dutch astronomer Van Woerkom had noticed in 1948 that there seemed to be about twenty times more short-period comets than he would have expected, the idea that short-period comets were really just ordinary comets from the Oort Cloud which had the misfortune to be captured by Jupiter was accepted for a number of years. However, as more and more comets were discovered, it became clear that something was wrong. The observed population of short-period comets was too large to be explained by the effects of Jupiter's gravity on comets from the Oort Cloud. The developing problem was twofold. Firstly, the capture of an individual comet by Jupiter is a very unlikely event. Only if a comet flies quite close to Jupiter can enough gravitational energy be exchanged to slow the comet down and trap it in the inner solar system. With comets arriving from all directions, including from above and below the plane of the planets, the chances of crossing Jupiter's orbit just when the planet happens to be there, and to do so close enough to the plane of the solar system to be captured, are very small. The probability of capture in this way is so low it has been compared with the likelihood of hitting a bird with a single bullet fired into the sky at random. Also, most of the short-period comets are in low-inclination orbits and most go around the Sun in the same sense as the rest of the planets. Since comets from the Oort Cloud approach the Sun at all angles to the ecliptic plane, including in orbits that go around the Sun in the opposite direction to the planets, it was puzzling that orbits of the short-period comets were not more randomly distributed.

In 1972, physicist–astronomer Edgar Everhart tried to resolve this problem. He suggested that the short-period comets were derived not from the capture of just any Oort Cloud comet, but rather from a subset of such comets which had specific orbital characteristics that made them likely to be captured. In particular, Everhart suggested that the short-period comets were Oort Cloud comets which had entered the zone of 4–6 AU from the Sun close to the plane of the solar system. Everhart's model could explain why the short-period comets population looked the way it did, but it was not long before another problem showed up. The following year Paul Joss from Princeton University looked at Everhart's model and put in some estimates for the capture rate, the likely lifetime of a typical short-period comet, and so on. In just two pages of text he showed that there were hundreds of times too many short-period comets to be explained by Everhart's capture model. Of course, not everyone agreed with him, but it did look as if something was missing from the equation. While the dynamicists pondered this problem, a new piece of the puzzle was about to be turned up by a strictly observational astronomer.

The Centaurs

The 1970s was a golden age for solar system exploration. Robot explorers orbited Mars, landed on Venus and flew sunwards to photograph the innermost planet, Mercury. In 1976, two Viking spacecraft landed on Mars to search for signs of life and, as the decade drew to a close, Voyagers 1 and 2 visited the gas giant planets Jupiter and Saturn. As mission followed mission, thousands of stunning images flooded back to Earth. One by one a dozen or more planets and moons were transformed from blurry images or points of light into individual worlds with distinct personalities. Across the solar system, newly discovered mountains, valleys and craters were mapped, catalogued and then named. Almost overnight planetary science moved from the realm of astronomy to become more akin to geology and geography. However, despite the huge amounts of data being returned from space, a few areas of solar system research remained the province of traditional ground-based telescopic observers. With space missions concentrating on exploring the planets and their satellites, the rest of the solar system seemed to be something of a backwater. The numerous comets and asteroids lacked glamour and received comparatively little attention. None the less, a tiny band of astronomers struggling to make sense of the population of small bodies was beginning to make some progress in understanding the structure and composition of the main asteroid belt. As the Voyager missions revealed that much of the outer solar system was dominated by icy material and that large impact craters were found throughout the solar system, a few people began to ask questions about what else might lurk in the huge volume of space beyond Jupiter.

Although detailed thinking about the trans-Neptunian region had hardly begun at this time, the idea that a population of hitherto unknown planetesimals might exist received a boost in 1977 with the discovery of an unusual new member of the solar system. The object was found by Charles Kowal, a native of Buffalo, New York State. Always interested in an astronomical career, Kowal moved to California when he was sixteen and took a degree in astronomy at the University of Southern California. Tired of working 40 hours a week to support himself while studying, he decided not to stay on to gain a higher degree and, as the end of his studies approached, he wrote to various observatories in search of a job. After considering several possibilities he eventually accepted a position as a research assistant at the Mount Wilson and Mount Palomar Observatories, then home of some of the world's biggest telescopes. During this period Kowal worked for such well-known astronomers as Allan Sandage and Fritz Zwicky. He became interested in the solar system because asteroid trails frequently appeared by chance on the photographs of distant galaxies which he was taking for Zwicky. Around 1970, Kowal started searching for asteroids whose orbits brought them close to the Earth and he soon realised that the 1.15 m Schmidt telescope on Mt Palomar, which he had used during observations for Fritz Zwicky's projects, would be an ideal tool to search for new satellites of the outer planets. The Schmidt telescope, essentially a huge camera, was capable of taking photographs covering an area of sky 6 degrees × 6 degrees and a single plate would encompass the whole satellite system of a planet like Jupiter. Using this telescope, Kowal discovered Jupiter's thirteenth satellite in 1974. He named the new satellite Leda. In 1976, he expanded his work to begin a systematic photographic survey aimed at detecting hitherto unknown distant solar system objects.

For his search, Kowal used a photographic blinking technique similar to the one used by Clyde Tombaugh four decades earlier. Like Tombaugh, Kowal took a series of exposures concentrated on positions spread along the ecliptic and then blinked them to search for moving objects. On 1st November 1977 he was examining a pair of plates which he had exposed two weeks earlier when he found a faint object whose slow motion across the sky, equal to about 3 arcminutes per day, suggested that it might be about as distant as the planet Uranus. Following the established convention, Kowal reported his find to the International Astronomical Union's (IAU) Central Bureau for Astronomical Telegrams.

The Central Bureau for Astronomical Telegrams, known sometimes as simply CBAT and occasionally by the rather more Orwellian

name of the 'Central Bureau' is housed at the Smithsonian
Astrophysical Observatory in Cambridge, Massachusetts. Its director
was Brian Marsden, an avuncular, expatriate Englishman who had
developed an interest in cometary orbits as a schoolboy in the 1950s.
After getting a degree from Oxford University, Marsden left England
for Yale University and enrolled in that establishment's respected
celestial mechanics programme. He was awarded a PhD in 1965 and, at
the invitation of Fred Whipple, moved to the Smithsonian
Astrophysical Observatory. Three years later he became director of
the Central Bureau, responsible for the issuing of IAU telegrams and
circulars. In the days before the internet, IAU telegrams were used to
alert astronomers to potentially interesting discoveries such as
comets, variable stars, supernovae and other rapidly changing phe-
nomena. Today, the telegram service has been retired in favour of
electronic mail, although printed copies of announcements still go
out by regular mail as postcard-sized IAU circulars.

Marsden shared the news of Kowal's find with the Minor Planet
Center, a sort of central clearing house for observations of comets and
asteroids. Originally, centres for the calculation of asteroid and comet

Figure 2.1 Charles Kowal at the
blink comparator which he used to
discover the first Centaur, Chiron.
(California Institute of Technology.)

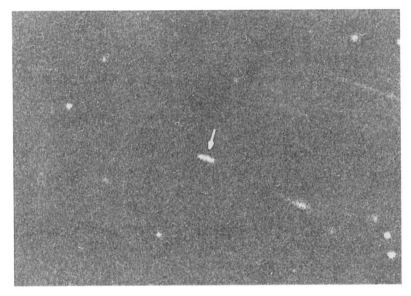

Figure 2.2 Charles Kowal's discovery image of Centaur 2060 Chiron. The trail is much shorter than the fainter one of a main belt asteroid at the upper right edge. This indicates that Chiron is moving more slowly, and so is more distant, than a typical asteroid. (Charles Kowal.)

orbits had been concentrated in Europe, with much of the work being done in Germany. Predictions of the positions of known asteroids appeared in the annual publication *Kleine Planeten* and newly discovered objects were announced via the circulars of the Rechen-Institut. These arrangements collapsed completely during the Second World War. After the war the *Kleine Planeten* was taken over by the Institute of Theoretical Astronomy in Leningrad and became the annual publication *Efemeridy Malyth Planet*. The work of the Rechen-Institut was transferred to a new Minor Planet Center established in 1947 by Paul Herget in Cincinnati, Ohio. The purpose of the Minor Planet Center was to receive observations and publish circulars containing accurate positions of newly discovered comets and asteroids. The center remained in Cincinnati until Herget's retirement in 1978, at which time it moved to Cambridge, Massachusetts, and came under Marsden's wing. Until he retired in August 2000, Marsden headed a small staff of experts in the esoteric field of orbital mechanics. The group, now headed by Dan Green, record observations of solar system objects and circulate the details to other interested parties, so that new sightings can be confirmed and followed up in a timely fashion.

Once sufficient observations have been made, the staff of the

Minor Planet Center calculate and then publish the orbits of the newly discovered bodies. At this point the Director of the Minor Planet Center has the sometimes delicate task of attributing credit for the discovery. Comets are usually named after their discoverer, or discoverers, although there are a few exceptions to this rule. Comets Halley and Crommelin are named after the individuals who calculated the details of their orbits rather than the astronomers who first observed them, and comets which are discovered using satellites or automated search telescopes are named after the project, rather than the people involved. Thus there are numerous comets called Solwind, IRAS, SOHO and LINEAR. Unlike comets, asteroids initially receive a temporary designation until their orbits have been calculated with sufficient precision that they can be found at any time in the foreseeable future. Once this has been done, the object is assigned a permanent minor planet number and the discoverer is invited to suggest a name for the new object.

Since photographs of Kowal's new object showed no sign of a comet-like coma, it was given the provisional asteroid designation of 1977 UB. The designation followed a simple code in use since 1925 which provides information on the approximate time of discovery. This designation is used until a permanent name and number can be assigned.[†] The first part of the designation is simply the year of discovery and the second part defines when in the year the object was first seen. The first half of the first month of the year is designated A, the second half B and so on throughout the alphabet, ignoring the letters I and Z. The second letter is assigned in the order that reports are processed by the Minor Planet Center. So the first object reported in the first half of January 1977 was known as 1977 AA, the second 1977 AB and so on until the middle of the month when the designations 1977 BA, 1977 BB, 1977 BC started to be applied. Kowal's object was the second one reported in the second half of October and so it was designated 1977 UB.

Luckily, another astronomer, Tom Gehrels, had observed the same area of sky on the 11th and 12th of October. An examination of his photographs soon revealed the new asteroid near the corner of his photographs. With these extra positions, together with other

---

[†] Until 1925 there had been three previous cycles (1893–1907, 1907–1916 and 1916–1924) during which pairs of letters were assigned sequentially without regard to the time of year.

observations made a few days later, it was possible for Brian Marsden to calculate rough details of the new object's orbit. These showed that 1977 UB was indeed distant from the Sun, but that it did not appear to be a comet entering the solar system for the first time. The initial calculations suggested that it was in a roughly circular orbit between the planets Saturn and Uranus. If this was true, then it represented a completely new class of solar system object and there was much interest in trying to determine the orbit more accurately. To do this, astronomers used early estimates of the orbit to project the object's motion back in time. They then hunted through their libraries of photographs to see if there was any chance that it might have been recorded accidentally on images taken for some other purpose. Since the new object was fairly bright, and so quite easy to find once one knew roughly where to look, Kowal was able to find images on two photographs he had taken in September 1969. With this new information, more images were soon found, some from 1976 and some from August 1952. A new orbit derived from these observations allowed William Liller to locate the object on a plate taken in 1941 with a 61 cm telescope at the Boyden Observatory in South Africa. A number of other images were also found, including ones from 1943, 1945 and 1948. There was even one on a plate taken in the USA in 1895 during predelivery testing of the Boyden Observatory telescope.

When an object is located based on a projection of its orbit into the future, the observation is called a recovery so, following the rather diabolical practice of film makers and authors in producing prequels to popular films and novels, observations made by projecting an orbit backwards in time are called precoveries. Searches for precoveries are quite common when interesting moving objects are discovered. This is because old observations can be very useful in refining the object's orbit, particularly when the object is very distant from the Sun and so moving only slowly across the sky. When observations span only a small period of time, the object's apparent motion, known as its observed arc, will be small. When observations covering this arc are used to calculate the object's orbital parameters quite large errors can result. By providing a much longer baseline of observations, or a longer arc, old observations make possible a more precise determination of the orbit, which in turn allows better calculations to be made of the object's past and future positions. These improved estimates of its past position sometimes make further precoveries possible, and this improves the knowledge of the orbit still further. In the case of

1977 UB the precoveries soon covered a timebase of over 80 years. This provided a big enough arc that its orbit was soon defined well enough that it could be numbered minor planet 2060. At this point, the Minor Planet Center, which acts under the authority of the International Astronomical Union, asked Kowal to give the new asteroid a name.

The tradition of asteroid naming is a long one. It began in the early 1800s, when the Italian astronomer Piazzi named the first asteroid Ceres. The next two hundred or so asteroids also received classical names, but with the introduction of photographic techniques for asteroid hunting in the 1890s, and a consequent surge in new discoveries, the naming protocol began to be relaxed. Soon a variety of people, places and organisations started to find a place in the heavens. In the case of obscure main belt asteroids the names chosen can sometimes be quite frivolous. Minor Planet 2309 Mr Spock, which is named after a cat, who was in turn named after the TV character, is a legendary example. Fortunately, Kowal, who describes himself as 'One of those old fashioned people who think that asteroids should be named carefully' and who perhaps suspected that there was more to 1977 UB than met the eye, was more circumspect. Since 1977 UB had been found early in his survey, and hopeful of finding some more objects later, Kowal looked for a group of mythological characters unrepresented amongst the asteroids. He found that the Centaurs, strange creatures that were half human and half horse fitted the bill. From dozens of Centaurs named in ancient literature, Kowal chose the name Chiron (pronounced Kai-ron), the most prominent Centaur and arguably the one with the best reputation. Mythologically speaking the Centaurs were a rowdy bunch given to drinking, rape and pillage, but Chiron was said to have devoted his efforts to astrology, medicine and the arts. He is described by some scholars as the King of the Centaurs and was known as a teacher as well as a healer. The choice was highly appropriate because the object was in an orbit between Saturn and Uranus and mythologically speaking, Chiron was the son of Saturn and grandson of Uranus. Although Kowal could not have known it at the time, it was an inspired choice.

The discovery of 2060 Chiron raised several interesting questions which occupied the popular media until the initial excitement died down. One question that was asked, and that we shall encounter again later in a different context, was whether Chiron was a new planet. Although the initial observations did not reveal anything about Chiron's surface, making it impossible to be sure if it was covered in reflective material like ice or darker material such as rock and dust, it was clear from its faintness that Chiron could not be more than a couple of hundred kilometres across. Since there are a number of main belt asteroids bigger than this, it was clear that Chiron was not a planet in the traditional sense of the word. However, classifying Chiron as a minor planet begged the question of whether it was unique. Was it just the brightest member of a new trans-Saturnian asteroid belt? Soon, further calculations based on the larger arcs obtained by using the precovery observations were available. These showed that the orbit of Chiron was not the circle originally estimated, but was an ellipse ranging from inside Saturn's orbit to a point just inside the orbit of Uranus. Such an orbit is not stable and it was clear that Chiron could only remain a denizen of the Saturn–Uranus region for, at most, a few million years. Although neither its precise past nor its eventual future could be calculated, it was easy to show that Chiron will eventually approach close to either Saturn or Uranus. When this happens gravity will drastically change its orbit, either moving it further into space or perhaps sending it closer to the Sun. Whatever its ultimate fate, Chiron is only a temporary resident of the outer solar system.

Around the time of its discovery, it was variously suggested that Chiron could have been an escaped satellite of one of the outer planets, a rocky asteroid somehow ejected from the main asteroid belt or perhaps a giant comet. The comet theory drew a parallel with another unusual asteroid, 944 Hidalgo, which although rather smaller than Chiron was also in an unusually eccentric orbit. Hidalgo was thought by some astronomers to be a comet from which all the water ice had been removed and which as a consequence was no longer active. However, since in its present orbit Chiron does not approach the Sun closely enough to sublime any water ice on its surface, Marsden and co-workers suggested that Chiron was not so much a dead comet, but rather one which had never lived. If they were right and Chiron was a comet, it was a big one; its brightness suggested that it was about fifteen times bigger than the nucleus of Halley's comet.

A few more details of the nature of Chiron were revealed a decade later when David Tholen of the University of Hawaii, amongst others, showed that Chiron was brightening faster than expected as it approached the Sun. Like planets, asteroids do not shine by themselves, they merely reflect sunlight and their brightness at any given time depends on a number of factors. The main ones are the size of the object, how reflective it is and its distance from both the Sun and the Earth. Although it was not clear how reflective Chiron was, and therefore what its absolute brightness should be, it was fairly easy to calculate how its brightness should change as it moved around the Sun. Tholen and co-workers Dale Cruikshank and William Hartmann found that Chiron was not sticking to the rules; throughout the late 1980s it was getting too bright to be explained by just its steadily decreasing distance from the Sun. An airless object's reflectivity, or albedo, is determined by its surface composition and this is unlikely to change dramatically over a period of a few years, so the most obvious explanation was that the extra brightening was due to Chiron developing a comet-like coma of gas and dust. A coma would drastically increase the area of material reflecting sunlight and could cause the anomalous brightening. These speculations were confirmed in 1989 when Karen Meech and others took images which showed that Chiron had indeed developed a coma, and even had a comet-like tail of material blowing away from the Sun.

At first, this cometary outburst was ascribed to Chiron warming up as it approached the Sun. It was speculated that solar heating somehow caused gases trapped below the surface to blow a hole in an insulating crust and allow a cloud of gas and dust to escape. Although water ice would be expected to remain frozen at the distance of Chiron, other volatile gases such as carbon monoxide, nitrogen and methane could be responsible. These gases can be trapped in amorphous (water) ice and can be released if the ice undergoes a change to the more regular, and familiar, crystalline form of ice. This transition from amorphous to crystalline ice can occur at quite low temperatures and has been proposed to explain the activity of various comets at great distances from the Sun. However, a careful study of old photographs showed that Chiron's activity was not restricted to periods when it was relatively close to the Sun. Outbursts were detected even when Chiron was near the most distant point of its orbit. To confuse things still further, Chiron's activity did not continue as it approached the Sun and actually diminished or even stopped during its perihelion passage in the 1990s. Whatever the source of the outbursts, it seemed that Chiron looked like an asteroid some of the time, but like a comet the rest of the time. Just like the mythical Centaurs, it was neither one thing nor the other.

Charles Kowal continued his search for about eight years, finally finishing in February 1985. By then he had observed 160 fields totalling 6400 square degrees of sky. Although he found four comets and several Earth-approaching asteroids, Kowal never did find another Centaur. Chiron remained a lonely enigma for almost fifteen years. The next step forward came from a project initiated by Tom Gehrels, who had himself unknowingly recorded Chiron in 1977. The Spacewatch project was established by Gehrels to make a long-term, systematic search for new solar system objects using electronic detectors instead of photographic plates, and computers instead of blink comparators. From a telescope on Kitt Peak in Arizona, Spacewatch was scanning the skies for several nights a month, making repeated observations of the same area to find objects that moved noticeably in a few hours. Most of Spacewatch's discoveries had been of asteroids

Figure 2.3 An image of Chiron taken with a CCD camera in the 1980s. The telescope has been tracked to allow for Chiron's motion, making the stars appear as streaks. Despite this, Chiron does not appear pointlike; the faint haze around it is evidence for a cometary coma. Images like this confirmed that Chiron is a giant cometary nucleus. (Karen Meech.)

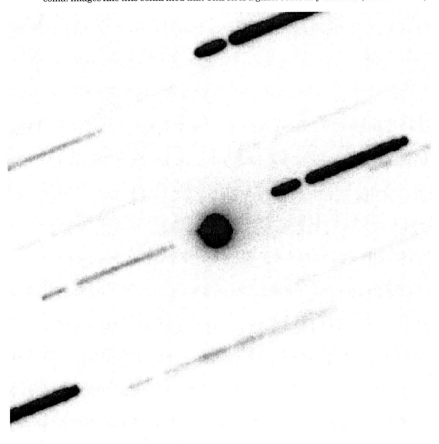

close to the Earth but, on 9th January 1992, David Rabinowitz was in the Spacewatch control room when the system's moving-object detection software drew his attention to an object apparently moving too slowly to be a normal asteroid.

Rabinowitz knew that false alarms resembling faint slow-moving objects were quite common. From time to time, the software linked together marginal detections of stars and electronic noise and mistook them for a single object moving across the images. However, Rabinowitz immediately realised that this source was much brighter than a typical false alarm. A quick examination of the images showed that the new object was pointlike, confirming its likely reality and suggesting that it could be a new, distant asteroid. Rabinowitz phoned Beatrice Mueller, who was working at the nearby 2 m telescope and she immediately agreed to try and observe the new object. She made the observations the same night. Within a few days, further observations had been made by the Spacewatch telescope and additional sets of positions were being reported by other astronomers.

One observation was from a pair of plates which had been taken on New Year's Eve by the comet-hunting team of Gene and Carolyn Shoemaker and their colleague David Levy. This group was really searching for fast-moving objects, but on this night their photographs also contained something that was moving rather slowly. They only had observations on a single night, not enough for the Minor Planet Center to do much more than file the positions away for future reference and certainly not sufficient for the object to be recorded as a possible new discovery. Because of this Carolyn Shoemaker admits she did not give measuring the object's position very high priority. At the time she thought it was probably a more-or-less ordinary asteroid, unlikely to be followed up by anyone else and so probably fated to be lost again. The observations were eventually reported to the Minor Planet Center on the 13th of January. They arrived about the same time as reports of another independent discovery made from photographic plates taken on 9th and 10th of January by Eleanor 'Glo' Helin. Well known for her work on asteroids passing close by the Earth, Helin was observing from Mt Palomar as part of a long-running search for fast-moving objects. As usual, while at the telescope she was concentrating on scanning her photographs as soon as possible so that any fast-moving objects which turned up could be reported and followed up immediately. Only once the observing run was over, and she was back in her office in Pasadena, California, did she have time to return to the photographs and search for anything moving slowly. Although the limited capability of the Palomar Schmidt telescope meant that very faint objects were likely to avoid detection, this time there was something recorded on the film.

Although the Spacewatch team had been first to make a report of their detection, all the sets of positions were published together and the new object was given the temporary designation 1992 AD. From the preliminary orbit calculated using the January observations, it was possible to begin a search for earlier detections. Soon a candidate was found on a plate exposed a year earlier by the Shoemakers. Assuming this sighting was indeed 1992 AD, another detection was found by Beatrice Mueller on a 1989 image. Soon 1992 AD turned up on photographs taken in 1982 and 1977. From all these observations a definitive orbit could be calculated and the new object received the minor planet number 5145. Now it needed a name. David Rabinowitz was interested in moving away from the tradition of naming asteroids after characters from Roman or Greek mythology, especially since he felt that the Centaurs as a group were an unsavoury bunch. He favoured naming outer planet asteroids after creatures from the creation myths of a number of different cultures. He suggested that the new object be called Chaos. This would have been a very appropriate name for an object in a planet-crossing, and so probably unstable orbit. However, tradition prevailed and the object was eventually named 5145 Pholus, who was Chiron's brother.

Like that of Chiron, the orbit of Pholus is unstable over a timescale of 10–100 million years, but there the similarity ended. Soon after Chiron had been discovered, astronomers had determined that its surface is neutral in colour. This is to say that it reflects all wavelengths equally and the light that comes back from it looks almost the same as the sunlight which arrives there. Pholus was very different. Within weeks of its discovery no less than three groups reported that the new asteroid was astonishingly red. Although Mueller mentioned it to Dave Rabinowitz, probably the first to comment on this in print was David Tholen from Hawaii. Tholen noted in an IAU Circular that Pholus was the reddest asteroid he had ever observed. So unusual were these colours, that Beatrice Mueller and her co-workers entitled the paper describing their findings 'Extraordinary colors of asteroidal object 1992 AD'. Pholus, so much like Chiron in terms of its orbit, seemed to be completely different physically. The first suggestion as to why Pholus was so red was that its surface was coated with a layer, not necessarily all that deep, of carbon-bearing molecules formed by the action of cosmic radiation on a surface that originally comprised mainly simple ices. If this was true, then it would suggest that Pholus' surface was older than Chiron's. Exactly how much older, no-one could say.

This conclusion is of such importance to our story that it is worth spending a few moments to see how it was reached. Astronomers use a number of techniques to study asteroids and one of these is called photometry. In photometry, the light arriving from an object of interest is passed through a filter onto a detector (usually some sort of electronic device) which can record the amount of energy received. The amount of energy from the source is then compared with that from a star of known brightness which has been observed with the same equipment under the same conditions. The ratio of the two values is described in logarithmic units called magnitudes. Five magnitudes corresponds to a brightness ratio of 100 and ten magnitudes to a factor of 10 000. Photometry is used in all areas of astrophysics, but here we need only concern ourselves with studies of asteroids.

If repeated photometric observations of the same object are made over a period of a few hours or days, then it is possible to determine if the object is varying in brightness. If this reveals a regular variation, usually called a lightcurve, then the object is probably irregularly shaped and is rotating, presenting different faces to the observer as it turns. Lightcurves of small- to medium-sized asteroids often involve changes of a few tenths of a magnitude over a period of a few hours. For reasonably bright objects, lightcurve observations are quite easy to do with even a relatively small telescope. Taking things a little further, by making measurements in a number of different filters, each of which pass only a narrow range of wavelengths, a sort of crude spectral fingerprint can be obtained. This tells the astronomer if the object reflects more blue light than red, or red light than blue, or if it reflects all colours equally. The filter system at most observatories uses five colour filters called U (Ultraviolet), B (Blue), V (Visible – about yellow), R (Red) and I (Infrared). Subtracting the magnitude in one filter from another gives what astronomers call a colour, for example U–B or B–V. It had been known for some years that if the U–B and B–V colours of asteroids are plotted on what is logically enough called a two-colour diagram they are not scattered about randomly. Instead, groups of objects cluster in specific regions of the diagram. These different colours are ascribed to the presence of different minerals, each of which reflects light in a slightly different way. Thus filter photometry is a popular tool of asteroid astronomers. The reason for its popularity is that it is fairly quick and can be applied to quite faint objects, so large numbers of asteroids can be observed and classified. This helps pick out the unusual and interesting objects from amongst the thousands of more-or-less ordinary ones. In this case, Pholus stood out at once because of its very red V–R and V–I colours.

Filter photometry is a good general tool, but provided that the objects are bright enough a better way of finding out about the composition of astronomical objects is to take spectra. This involves using a prism or a diffraction grating to spread the object's light out smoothly over a range of wavelengths. Spectroscopy is used extensively in the study of stars and galaxies since the spectra of hot objects often contain narrow lines which can be used to identify specific chemical elements and the physical conditions under which they exist. Spectroscopy of asteroids is not so simple since any features tend to come from molecules or minerals and are broader and much shallower than atomic lines. This makes them harder to detect so spectroscopic observations have to be concentrated on the brighter asteroids. Fortunately, both Chiron and Pholus are quite bright and both could be observed spectroscopically.

The spectra of Chiron and Pholus in the region between about 0.4 and 1 microns, which roughly correspond to the UBVRI filters, are featureless. The optical spectra reveal nothing that might provide a clue as to either object's composition. However, many simple molecules, especially those containing carbon atoms, have spectral features in the near infrared region of the spectrum. This region corresponds to wavelengths from about 1 to 4 microns and for a variety of technical reasons these wavelengths are much harder to observe than optical wavelengths. Luckily, at about the time Pholus was discovered, a new generation of infrared spectrographs was being put into service at several of the world's major observatories. One of these instruments, the then new CGS4 spectrograph at the UK Infrared Telescope (UKIRT) in Hawaii, was turned onto Pholus by staff astronomer Gillian Wright in 1992. Ironically, Gillian Wright's main interests were in extra-galactic astronomy, about as far as one can get from studying solar system objects, but as the scientist responsible for the CGS4 spectrograph she was just looking for a suitable target for some tests. She agreed to observe Pholus for a few minutes when it was suggested as a possible target.[†] The observations were very successful. They revealed that Pholus had infrared spectral features unique in the solar system, notably an absorption in the spectrum around a wavelength of 2.25 microns. This feature could not be identified with any certainty, but it was similar to features seen in chemical mixtures called tholins produced during laboratory experiments attempting to duplicate conditions in the early solar system.

The first tholins were produced by taking flasks of simple gases which were expected to exist in the atmospheres of gas giant planets and subjecting them to electrical discharges to simulate the effects of lightning on the primitive atmospheres. The results of these experiments were a mish-mash of carbon bearing molecules such as amino acids and a red-brownish material that was quite stable and difficult to destroy. This residue was named a tholin (from a Greek word meaning dirty) and will be familiar to anyone who has tried to clean up old glassware which has been used in organic chemistry laboratories. The presence of tholin-like absorption features in the spectrum of Pholus was confirmed when Dave Jewitt and Jane Luu of the University of Hawaii repeated the observations a year later and, thanks to technical improvements within CGS4 since the 1992 observations, obtained a much higher-quality spectrum. Jewitt and Luu also observed Chiron with CGS4, but they found no sign of spectral features similar to those in Pholus.

A number of other Centaurs have since been discovered, but many of them are very faint and not much is known about them. The best observed are 7066 Nessus (1993 HA_2), 8504 Asbolus (1995 GO) and 10199 Chariklo (1997 CU_{26}), all of which were discovered by the Spacewatch team. Chariklo is quite bright and photometry of this object shows that it is redder than Chiron, but less red than Pholus. Spectra from both UKIRT and larger telescopes show that Chariklo seems to show features due to water ice, but none of the Centaurs yet observed has the deep 2.25 micron feature seen on Pholus. Paradoxically, ice is not seen in most published spectra of 8504 Asbolus,[‡] which looks otherwise rather similar to Chariklo. However, there may be more to these spectra than meets the eye. Spectra of Chiron from the mid 1990s did not reveal the presence of ice there either, but more recent

---

[†] The suggestion came from me, I happened to be visiting Hawaii and was sitting in on the weekly UKIRT schedule planning meeting. It was a chance encounter that changed the direction of my scientific research programme. JKD.

[‡] But see page 119.

observations have detected weak ice features. Chiron is known to show comet-like activity and it seems that the ice has only become visible since a period of activity a few years ago. This suggests that the ice was always there, but that its spectral features may be easier to detect under certain conditions than others. Small changes in the structure or composition of their surfaces may explain why some Centaurs do show evidence for ice and others do not. In all probability they are all icy objects.

In 1998, Dale Cruikshank of the NASA Ames Research Center at Moffett Field near San Francisco combined a number of observations of Pholus with calculations of the spectra expected from various ices and molecules to make a model of its surface. He believes that the spectrum of Pholus can be explained by a combination of water ice, a dark sooty material, some mineral dust and some ices of a simple carbon-bearing molecule like alcohol. By adding these together in an appropriate way, his group can match the observed spectrum of Pholus quite well. Interestingly, although of course it is not a coinci-

Figure 2.4 The first published infrared spectrum of the Centaur 5145 Pholus. Taken by Gillian Wright from the UK Infrared Telescope UKIRT in 1992, it shows unusual spectral features in the 2 micron region. The gap between 1.85 and 2.05 microns was not covered by this observation. (Originally published in *Icarus*, vol. 102, p. 67. Courtesy John Davies/Academic Press.)

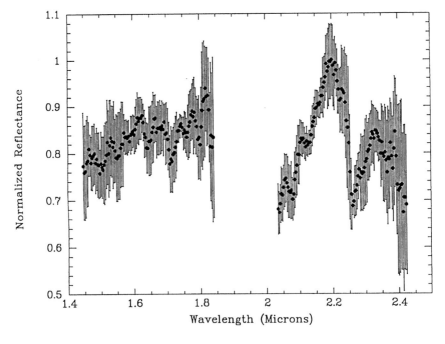

dence, the materials chosen by Cruikshank for his model match pretty well with our current understanding of what makes up the nucleus of a comet.

Cruikshank's model offers a possible explanation of the dichotomy between Pholus and Chiron. Perhaps they are essentially similar objects, being giant comet nuclei composed of mostly primitive solar system ices and some dust. They were probably formed about the same time as the planets and have been stored in deep freeze in the outer solar system since then. Recently, they were each moved inwards by some subtle combination of gravitational forces. Perhaps Pholus is still covered by an ancient crust formed when its ices, being bombarded by cosmic rays (high-energy particles from beyond the solar system) formed a layer of tholin-like material which helps to preserve the ices frozen beneath the surface. Chiron, while basically similar, is for some reason now subject to comet-like outbursts which bring fresh material, probably ices, from its interior. Geysers of material jetting out from cracks in the crust of Chiron may have fallen back, or recondensed, coating the surface with a lighter, neutral crust. This fresh material could be hiding any evidence for older material that might be present. Either that or the outbursts have simply blown away the ancient crust completely. Although plausible, if this explanation is true it is not clear why Pholus has not experienced similar activity.

Possibly either Chiron or Pholus was once nearer the Sun. If this is

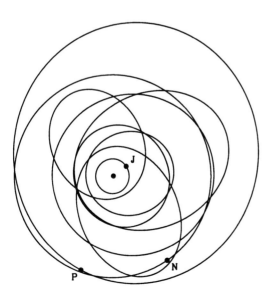

Figure 2.5 The orbits of six Centaurs compared with those of Jupiter, Neptune and Pluto. At aphelion the Centaurs are close to Neptune's orbit, suggesting that they might have originated in the trans-Neptunian region. In this and the following figures the orbits of the Earth and the other inner planets are not shown as they would be almost indistinguishable on this scale. (Chad Trujillo.)

so then either of them may have had its surface drastically altered before the random effects of the gravity of the inner planets ejected it outwards again. This is by no means impossible. Calculations made by Gerhart Hahn and Mark Bailey in 1990 suggest that there is a high probability that Chiron has once been in an orbit more like that of a short-period comet. If so, a close approach to the Sun could have stirred up a burst of activity and removed its ancient crust leaving behind a fresher and more active surface for us to observe today. On the other hand, perhaps it was Pholus that ventured briefly sunward and experienced some drastic chemistry on its surface, and its red colour today is just a bad case of cosmic sunburn. If the surface of Pholus has been burned away, or chemically reprocessed, then perhaps Chiron and the other Centaurs are more representative of primitive solar system material and the idea that Pholus has an ancient surface is just plain wrong. Unfortunately, the chaotic nature of the Centaurs' planet-crossing orbits makes it impossible to project their positions far enough back in time to find out.

The planet-crossing orbits of the Centaurs mean that they cannot remain in their present locations for even a tiny fraction of the age of the solar system. The fact that some, indeed an increasing number of them, are being detected means that the Centaur population is being continuously replenished. The discovery of the Centaurs lent credence to increasingly detailed suggestions that there was a reservoir of objects beyond Neptune that could be diverted into Centaur-like orbits and which would then evolve inwards to become short-period comets.

The mystery of the short-period comets

By the early 1970s the rate at which short-period comets were being discovered was increasing suspicions that there were too many to be explained by objects captured directly from the Oort Cloud. One of the people considering this problem was a professional physicist who had started out with an amateur interest in astronomy. As an amateur astronomer, Edgar Everhart discovered two comets, one in the summer of 1964 and another in 1966. Perhaps it was this that sparked his interest in the distribution of comets and led him to write his first scientific papers on the subject. His papers were well received and marked a transition from the world of physics into that of astronomy. Everhart assumed the directorship of the Chamberlin Observatory in Denver, Colorado, and developed an interest in astrometry, the measurement of the positions of objects in the sky. Everhart took photographs of faint comets and, in the basement of his home in the Colorado mountains, he measured their positions using a machine he had built himself.

Everhart became interested in orbital dynamics and started to apply new methods of numerical integration to the study of how comets' orbits evolve under the influence of the gravitational fields of the planets. Although he had shown in 1972 that captures by Jupiter were possible, it seemed to him that there were far too many short-period comets to be explained by Jupiter captures alone. In 1977, following earlier work done by the Russian dynamacist Kazimirchak-Polonskaya, he tried to solve the problem of the excess of short-period comets. His method was to consider Oort Cloud comets approaching the Sun on initial orbits that did not penetrate the solar system as far as Jupiter, but which just skimmed the edge of the planetary region in the vicinity of Neptune. The gravitational effect on a comet during an encounter with Neptune is about 11 000 times less than the effect of an encounter with Jupiter, but Everhart reasoned that Neptune does not have to do all the work. All that was needed was for Neptune to change the comet's orbit enough to move its perihelion inwards, so that it has a chance of falling under the gravitational influence of Uranus, the next planet in. If this can be done, then an encounter with Uranus may move the comet inwards again until it falls under the gravitational control of Saturn. The gravity of Saturn can then direct the comet into an orbit which passes close enough to Jupiter for it to be captured into a short-period orbit.

Everhart simulated this process by examining the evolution of thousands of cometary orbits on a computer using what astronomers call 'Monte-Carlo' methods. These are numerical techniques that take their name from the famous casino town where small balls bounce around roulette wheels and finish up in random slots around the edge. The computing technology of the 1970s did not allow a detailed calculation of exactly how the orbits of each hypothetical comet would evolve, so Everhart was forced to use a number of shortcuts. These reduced the amount of computing time by a factor of 500, but even so it required many hours to carry out the simulations. However, although it worked in the sense that Everhart's calculations showed that short-period comets could be created in this way, the calculations also showed that it was a very inefficient process. Many, many comets were ejected from the solar system for each one that was captured.

The first of the simulations considered the diversion of comets towards Uranus by Neptune. These revealed that only 18 out of 12 230 simulated comets made the jump inwards. However, a second simulation showed that once a comet had been steered into Uranus' control, its chance of going on toward Saturn's sphere of influence was 40/69. Further numerical experiments showed that the likelihood of moving from Saturn's influence to Jupiter's was 229/500 and that, once there, the likelihood of becoming a short-period comet in the Jupiter family was 92/229. Since the final likelihood of making a short-period comet is reached by multiplying all of the above probabilities together, it was clear that only one in several thousand comets would be captured, the rest would be ejected from the solar system during one of the planetary encounters along the way. The process was also very slow. A typical comet that encroached Neptune's sphere of influence would survive over 40 million years before being sent inwards to Uranus. Indeed, the capture of Oort Cloud comets by Neptune could only resolve the problem of the excess of short-period comets if there was a very massive inner Oort Cloud containing perhaps 700 times the mass of the Earth. That such a huge mass of cometary material should exist seemed unlikely. Another mechanism to supply the short-period comets was needed.

A possible explanation came from Uruguaian astronomer Julio Fernandez. While a student at the University of Montevideo, Fernandez had become interested in the problems of the formation of the solar system, how the planets had formed, how the direction of planetary rotation was established and so on. However, in the late 1970s the political situation in his home country had become unstable. Indeed, the mood throughout much of South America was generally unfriendly towards science and other cultural activities. Feeling uncomfortable with this environment, Fernandez moved to Spain. He spent a year at the National Observatory at Madrid, where he continued to read and think about comets and their relevance to the formation of the solar system. Fernandez found the idea of a belt of comets beyond Neptune put forward by Kuiper and Whipple exciting. Before long he started to consider if such a belt could have anything to do with the apparent excess of short-period comets. Since the rate at which Oort Cloud comets could be captured into short-period orbits was rather controversial, he attacked the problem from another point of view. Using the results of people like Everhart, who had shown that the most common fate of a comet from the Oort Cloud entering the inner solar system was to be accelerated by planetary encounters until it escaped from the solar system forever, Fernandez estimated that for every comet finally captured into a short-period orbit, 600 or more would be ejected into interstellar space. By combining the number of known short-period comets and their expected lifetimes, he was able to estimate the rate at which new ones had to be captured to maintain a healthy comet population. From this, and assuming the 1:600 ratio of captures to losses was correct, he was then able to estimate the rate at which Oort comets were being ejected into interstellar space. The conclusion that he reached was that some 300 comets a year were being lost. Such a high rate of wastage could not be maintained over the age of the solar system unless there were an improbably large number of comets in the Oort Cloud to begin with. Fernandez concluded that there must be an alternative source for the short-period comets.

Having decided that short-period comets could not, in the main, be captured from the Oort Cloud, Fernandez tried to decide if they could diffuse inwards from a hypothetical trans-Neptunian comet belt. He took as a starting point that there was about one Earth mass of comets in a belt stretching from 35 to 50 AU. He assumed that this material was distributed as bodies with sizes ranging from a small comet nucleus up to something about the size of the Earth's moon. He then tried to estimate the effect of close encounters between objects in this belt. His objective was to see if the gravitational effects of the largest ones could perturb the orbits of the smaller ones into the planetary region. His simulations of thousands of random encounters between hypothetical trans-Neptunian comets soon indicated that a flattened disc of planetesimals orbiting just beyond Neptune could supply a couple of comets per year into the region of the outer planets. Since any objects sent inwards from this disc would start close to the plane of the solar system, their further evolution would be quite rapid in astronomical terms. Using Everhart's estimates that there was a roughly 50:50 chance that a comet crossing the orbit of one giant planet would be sent inwards to encounter the next planet along, Fernandez concluded that once a comet had moved into Neptune's sphere of influence it had about one chance in sixteen of evolving into a short-period comet. The fate of the remainder was to be ejected back outwards again. The important difference was that unlike the 1:600 estimate of captures direct from the Oort Cloud, a loss rate of sixteen to one did not require an unrealistic number of Oort Cloud comets to start with.

Having solved the problem to his satisfaction, Fernandez, who had by now moved to a position at the Max Planck Institute for Aeronomy in Lindau, Germany, wrote up his conclusions. He sent them for publication in the *Monthly Notices of the Royal Astronomical Society*, where Edgeworth's initial speculations had appeared three decades earlier. No-one at the Royal Astronomical Society seems to have commented on the omission of any references to Edgeworth's work and Fernandez's paper was published in 1980. At this point Fernandez felt there was no point in continuing to work on the origin of the short-period comets. He went back to working on issues associated with the formation of the planets and the dynamics and origin of the Oort Cloud. However, in closing his paper, he did pose the question of whether objects in this hypothetical belt could be observed. Using an estimate of the reflectivity of Pluto as a starting point, he concluded that objects in the comet belt might shine at about 17th to 18th magnitude. These would be difficult, although by no means impossible, to detect with late 1970s technology. We now know that while Fernandez was broadly correct in his conclusions about the existence of a trans-Neptunian comet belt, he was wrong about this last point. As we will see later, objects in the belt are much less reflective than Pluto and so they are much fainter than he thought.

A crucial advance in resolving the problem of the short-period comets came in the late 1980s with a series of papers from a group of theoretical astronomers based in Canada. A member of this group, and someone who would go on to play an important role in understanding the dynamics of the outer solar system, was physicist Martin Duncan. Born in London, England, in 1950 Duncan went, or rather was taken, to Canada when he was 18 months old. He grew up in Montreal. As a child, he always had an interest in how things worked and admits to driving his parents crazy by constantly asking typically small boy questions along the lines of, 'What would happen if ...?'. At school he enjoyed mathematics and when he discovered physics he thought it was a wonderful subject which would let him study how things really work. This enthusiasm led him to take a physics degree. Although he had never been an amateur astronomer and still finds it terribly embarrassing when people ask him to point out constellations in the night sky, it was during his degree that he developed an interest in astrophysics.

Martin Duncan graduated during a time of great excitement in astrophysics, with the discovery of exotic new types of objects such as quasars, pulsars and black holes. Caught up in this excitement, he went to the University of Texas at Austin to study for a PhD in general relativity. He started to work on the theory of how gravitational collapse forms black holes, but says he soon realised that solving this problem was going to take more than the time allocated to getting a single PhD. In the end, his thesis finished up being on the related subject of the orbits of stars around black holes. Like many young astronomers, he then had a fairly nomadic existence for a while. He worked at Cornell University in up-state New York before going back to Canada and spending five years in Toronto as an assistant professor at the university there. By then it was the mid 1980s and one of the hot topics in astronomy was the suggestion that an unseen companion star of the Sun, sometimes called Nemesis, might periodically inject comets into the inner solar system. It was speculated that the resulting comet shower would cause large numbers of impacts on the Earth, triggering mass extinctions like the one which removed most of the dinosaurs about 65 million years ago. Becoming interested in this possibility, Duncan started reading more about this topic. He soon realised that the injection of comets towards a very distant target, such as the Sun or the Earth, was very similar to the problem of studying the orbits of stars around a black hole. So, along with Scott Tremaine, the director of the Canadian Institute of Theoretical Astrophysics (CITA), a research establishment in Toronto, Duncan put together a proposal to study cometary dynamics. Having been awarded a grant to pursue this question, they hired a postdoctoral research worker called Tom Quinn to help with the project. This turned out to be a happy choice and led to a series of important papers on the dynamics of comets.

Advances in computing power since Everhart's work in the 1970s now made it possible to do more detailed calculations of the path of individual comets under the influence of the gravitational forces of the giant planets. In simulations described in their 1988 paper, 'The origin of the short-period comets', Duncan, Quinn and Tremaine followed the orbits of imaginary comets until they were either ejected

Figure 3.1 Martin Duncan, celestial mechanician. (Terence Dickinson.)

from the solar system completely or became visible to equally imaginary Earth-based astronomers. However, as Everhart had noted, the
diffusion of a comet into a short-period orbit can be a slow process. It
may take millions of orbits before the crucial interactions occur. So,
even with a new generation of faster computers, some approximations were still necessary to get the simulations done in a reasonable
time. One trick they used was to increase artificially the mass of each
of the giant planets by a factor of up to 40. This increased the rate at
which orbital evolution occurred in their mathematical model.
However, even with this shortcut, the simulations still took several
months of computer time to complete.

First, they tried to duplicate Everhart's conclusion that short-
period comets could be the result of the capture of Oort Cloud comets
with perihelia in a specially favourable region close to Jupiter
(4–6 AU). They created 5000 imaginary comets in their computers and
gave them a wide range of inclinations, allowing them to approach the
Sun from many directions. They then set the simulations running and
let the orbits evolve. As it turned out, they did not get the same result
as Everhart. They found just the opposite. Although comets entering
this region of the solar system could be captured, the population of
simulated 'cyber-comets' which was produced did not look anything
like the actual short-period comets observed by real astronomers. In
particular, their simulations more-or-less preserved the inclinations
of the original comets and did not produce a population of short-
period comets close to the plane of the solar system. Next, they investigated Everhart's other suggestion, that comets were captured by
Neptune and passed down from planet to planet until they reached
Jupiter. Once again, the resulting population did not look right; there
were too many comets in orbits with high inclinations compared with
the situation in real life. Their conclusion was that it was impossible
to form Jupiter-family comets in orbits close to the ecliptic plane if
they started with a population of Oort Cloud comets that could arrive
from all directions. The next step was inevitable, if the source was not
a sphere of comets, how about starting with a disc of material already
close to the ecliptic plane?

They set up their simulations again. This time they started with a
population of comets in orbits with inclinations of between zero and
eighteen degrees and on paths that already crossed the orbit of
Neptune. Once again they set the computers running and followed the
evolution of these cyber-comets as they either fell inwards towards

the Sun or were ejected into deep space. The result was dramatic. The simulations produced a population of short-period comets very much like the real thing. The small differences that still existed could easily be accounted for by their assumption that all comets coming within 1.5 AU of the Sun are discovered (which may not be the case for real comets) and by subtle effects caused by the need to increase the masses of the giant planets used in the simulations in order to speed things up.

Another conclusion of the study was that about 17% of comets that start in Neptune-crossing orbits finish up as short-period comets. This was three times more than Fernandez's estimate. While the disagreement of a factor of three was not huge (some physicists joke that the fundamental equation of astronomy is that 1 is approximately equal to 10), it implied that the number of comets in the comet belt could be correspondingly smaller. Only a fraction of an Earth mass of material was now required. Furthermore, this material could be injected into the Neptune-crossing orbits quite slowly, easily long enough for the disc to have survived for the age of the solar system. For the moment, it was not obvious exactly how material got from the comet belt into the Neptune-crossing orbits, but once it got there it was clear that it could supply the required number of short-period comets.

Duncan, Quinn and Tremaine went a few steps further in their calculations. They showed that Chiron, which had been discovered by Kowal a decade earlier, could well be a bright member of the parent comet population which was in the process of undergoing just such a diffusion inwards from what they called the 'Kuiper Belt'. Finally, they showed that for a reasonable size distribution, there ought to be more than a thousand objects in this Kuiper Belt that would have a magnitude of about 22, and so would be directly detectable with suitable ground-based telescopes. All that remained was to find some.

Shooting in the dark

Coincidentally, about the time that the existence of a trans-Neptunian disc was being demonstrated mathematically, came the discovery of an object that might be the first member of such a structure. In September 1992 two astronomers from the University of Hawaii's Institute for Astronomy reported the detection of a faint, slow-moving object orbiting the Sun beyond Pluto. The result took the community of solar system astronomers by storm, but for Dave Jewitt and Jane Luu it was not really a surprise. It was more a relief after five years of fruitless searching.

Dave Jewitt is a tall, thin Englishman with an acerbic wit and a fine sense of the ridiculous. Gold-rimmed glasses and rapidly thinning hair atop a wiry body create something of the illusion of a mad scientist, but he is a shrewd and determined astronomer. Born in Tottenham, North London, in 1958, he grew up in Enfield, another London suburb. Jewitt discovered astronomy at the age of seven. One evening in 1965 he was riding his bicycle home when he saw large numbers of meteors, many of them bright enough to shine through the haze of London street lights. Intrigued by this cosmic firework display, he began to look more closely at the stars. Soon he started to pick out some constellations and to learn his way around the night sky. Not long after, his grandparents gave him a telescope as a birthday present and with this he began to explore the craters on the Moon, the rings of Saturn and the satellites of Jupiter. After developing an interest in mathematics and physics at school, he decided to study for a degree in astronomy at University College, London.

While Jewitt was studying for his degree, spacecraft were returning some of the first good-quality images of the planets and revealing landscapes shaped by hitherto unknown craters and volcanos. Impressed by these discoveries, Jewitt thought he might like to pursue a career as a planetary geologist. Since planetary exploration was basically a NASA enterprise, he decided to do postgraduate work in the United States. On the advice of a professor at University College, London, he enquired about possible studentships at the University of Arizona and the California Institute of Technology (Caltech), both of which had strong planetary science research groups. He eventually settled on Caltech because, he says, their application form was a lot shorter and simpler to fill in.

Caltech is a large, famous and rich institution located on a campus in the city of Pasadena, to the north of Los Angeles. It controls the famous Mt Palomar observatory and also operates the Jet Propulsion Laboratory (JPL), the nerve centre for numerous NASA space missions. While studying at Caltech, Jewitt worked on data from the two Voyager spacecraft, which were returning stunning new data about Jupiter and its satellites. He also began to use the telescopes on Mt Palomar. During his trips to the mountain he learned about the reality of doing astronomy for a living and (like many other neophyte astronomers) had what he describes as many 'ugly experiences' there. However, in the process he learned how to do observational astronomy in a very direct, hands-on manner.

In 1982, while still a graduate student at Caltech, Jewitt set about recovering Halley's comet. Through his position at Caltech he had access to the large telescopes on Mt Palomar and says he thought, 'Gee, why not give it a try?'. The search was carried out on various occasions spread over several months, using a few hours here and there borrowed from other astronomers who had been assigned the 200 inch (4.8 m) telescope for the night. The comet was eventually found on the night of 16th October 1982, when Jewitt was observing with Ed Danielson. It was not an easy observation; there was a bright star very close to the predicted position of comet and the light from the star was blinding the detector. Jewitt found an old razor blade in his room and cracked it in half to make two knife edges. He put these into a little ring which he mounted in the focal plane of the telescope and used them to block off the bright star. The razor blade had obviously been used and had bits of stubble stuck on it. Jewitt and Danielson called them 'Hubble hairs' after the great Palomar astronomer Edwin Hubble. The recovery of the comet, the first time it had been seen for over 70 years, was reported on the front page of the

Los Angeles Times, bringing good publicity to the observatory and doing no harm to the career of Dave Jewitt.

Dave Jewitt's thesis work included optical and infrared studies of comets and he graduated with a PhD in 1983. That same year he left California and moved to the East coast of the United States, becoming an assistant professor at the Massachusetts Institute of Technology (MIT) in Boston. Here he continued to work in planetary astronomy. He made observations of comets and asteroids and published a series of scientific papers, including some on the activity of distant comets and the still-enigmatic Centaur Chiron. While at MIT he also began a programme to search for other small distant bodies because he found himself asking, 'Why does the outer solar system appear so empty?'. Having convinced himself that there must be something out there, he set out to find it.

Jewitt was joined in his search by a graduate student who had arrived at MIT by an even more convoluted route than his. Luu Le Hang was born in Vietnam and had come to the United States as a child in 1975, fleeing Saigon with her family as the North Vietnamese Army entered the capital. After settling in California, Luu Le Hang became Jane Luu and set out to get an American education. After learning to speak English and doing well in high school science classes, she entered Stanford University and graduated with a Bachelor's degree in physics. Although she had planned to continue her career by study-ing solid state physics, she was diverted into studying planetary astronomy by chance. Working at the Jet Propulsion Laboratory as a computer operator, where she had a routine job running computer pro-grams connected with NASA's Deep Space Network, she was sur-rounded by spectacular images of planets and moons which adorned the walls of the laboratory. Gradually these images seduced Jane Luu into beginning a PhD in planetary astronomy. She applied for, and was awarded, a graduate student position at MIT.

Dave Jewitt and Jane Luu did not work together directly at first. As part of her postgraduate studies, Luu had to complete two research projects, one of several qualifying requirements which had to be com-pleted before deciding on a subject for her PhD thesis. Her first project involved helping in a search for the optical counterparts of gamma ray bursts, but at some point Jewitt suggested that a search for outer solar system objects might make a good (if rather out of the ordinary) research project. Jane Luu recalls asking in effect 'Hasn't this been done already?'. Jewitt replied 'No'. She then asked 'Then why should

Figure 4.1 Jane Luu and Dave Jewitt. The picture was taken in the control room of the UKIRT telescope in 1994. (Jane Luu.)

we do it?' and was told 'Because if we don't, nobody will'. She agreed to give it a try.

They began their search with traditional technology, using Schmidt cameras and photographic plates in much the same way as Clyde Tombaugh and Charles Kowal. The plan was that they would first take a photograph of a region of sky, then return to that same patch of sky after a short time to take another photograph. Then they would compare the two images to see if anything had moved in the interval between the two exposures. Although they could expect to detect many moving sources, it was possible for them to have a good idea of how the very-distant solar system objects they were seeking would be moving across they sky. Since objects at the distance of Pluto take two to three hundred years to go around the Sun, they only move across the sky at a rate of about 1 degree per year. This is so slow that it would be very hard to detect were it not for the fact that the Earth is also moving around the Sun. Since the Earth moves comparatively quickly, it is effectively overtaking the distant object. So, provided you look in a direction that is almost directly away from the Sun, towards what astronomers call the opposition point, then distant objects appear to move backwards on the sky. This so called reflex motion is about the same for all distant solar system objects so there is a good chance that any objects behaving in this way are indeed in the outer solar system.

In the first half of 1987 Jewitt and Luu undertook several observing campaigns using the Schmidt telescopes at two of America's observatories, the Kitt Peak National Observatory in Arizona and the Cerro Tololo Inter-American Observatory in Chile. The Schmidt telescopes could cover a large region of sky, each recording images covering an area 5.2 degrees square in a single exposure. During each observing run they photographed a series of regions along the ecliptic, generally taking three images of each field. Each image was exposed for about one hour and the fields were rephotographed about a day apart, to give slow-moving objects enough time to move significantly between the observations. Although only two such images were strictly necessary, the third plate allowed the reality of any candidates to be checked, as any suspicious object seen on a pair of plates should have a counterpart with the same rate and direction of motion on the third.

The plates were examined during the week after the observing run. This was done at the headquarters of the US National Observatories, on Cherry Street in Tucson, Arizona, a few hours drive from the telescopes on Kitt Peak. The searching was done using a blink comparator installed in the basement of the building. According to Jewitt, the comparator was serviced by a technician who could never be found, but who would appear miraculously to fix problems before he could even be told about them and then vanish just as mysteriously afterwards. Each pair of plates took about 8 hours to examine, and most of them were searched independently by both Jewitt and Luu in order to check that nothing had been missed. The process was hard work and often excruciatingly dull. Jewitt recalls that it was difficult to maintain concentration for more than a couple of hours at a time. In the 1988 paper containing their results they describe the blinking process as 'placing considerable strain on the eyes' and it is worth considering in a little detail exactly what was involved. Each Schmidt telescope recorded its image onto a photographic plate 194 mm square, about the same area as the open pages of a paperback book. A main belt asteroid moves across about 1 degree of sky every four days, so during a one hour exposure it moves about 30 to 40 arcseconds. This movement corresponds to a trail less than half a millimetre long on one of Jewitt and Luu's plates. On a pair of plates taken a day apart the asteroid would appear as two such tiny trails, separated by about three quarters of a centimetre, rather like the two horizontal parts of the letters 't' in the word toot. More-distant objects, which were, of course, what they were looking for, move more slowly across the sky. They would appear as two streaks, each about a tenth of a millimetre long, just 2 or 3 millimetres apart. Every square centimetre of each plate had to be searched for evidence of such tiny marks, and every example found had to be noted and then re-examined under a microscope to check if it was real. Without exception, every candidate turned out to be false, created by either tiny defects in the photographic emulsion of the plate itself, or by specks of dust.

Luckily for Jewitt and Luu, and especially for their eyestrain, technology was moving on. By the late 1980s a new generation of electronic detectors called Charge Couple Devices (CCDs) were beginning to challenge the photographic plate as the best way to make images of the sky. Although when used with a suitable telescope, photographic plates can cover a large area, even the best ones are relatively insensitive. They convert only about 10% of the light which arrives into exposed grains of the film's emulsion. In contrast, electronic detectors can have very high efficiency, as much as 90% or more in some cases, but are rather restricted in size and so only cover small regions of the sky in a single exposure. Obviously, searching the sky efficiently needs a detector which is both very sensitive (or 'goes deep' as astronomers like to say) and yet which can cover a large region of the sky to increase the chance that there is actually something in the image to detect. There is a trade-off to be made between sensitivity and sky coverage. In 1987 it was not clear whether shallow and wide was better than narrow but deep, so Jewitt decided to hedge his bets and also try this new technology.

Today, CCDs form the backbone of almost every astronomical detector in use at optical telescopes. They are made from semiconductor materials, usually silicon, which have been treated, or doped, in such a way that when a photon of light falls onto the detector, it liberates an electron from the semiconductor. These electrons, being negatively charged, can be stored or moved about the device (which is often referred to as a chip) by varying electrical potentials applied across it. In operation, the CCD is mounted inside an instrument which is exposed to the sky through a shutter, just like the film in a normal camera. However, when photons arrive at the focal plane of the camera, they are not bound up in a chemical reaction with a film emulsion, but instead they release electrons into the thin wafer of semiconductor material. At the start of each exposure the CCD is set up with some fixed voltages, so that any electrons liberated by the incoming light stay put, and hot spots of electric charge build up within the chip wherever light falls on it. Once the exposure has been completed the shutter is closed. Then, by applying a series of changing voltages across the chip, the electrons are read out in a systematic manner which preserves a record of where on the chip the charge came from. In essence, the chip is divided into a grid of tiny squares called picture elements, or pixels, and the charge in each pixel is recorded during the readout process. Once the readout is complete, the voltages across the chip are reset to erase all evidence of what happened during the previous exposure. Finally, the shutter is opened again and the whole process is repeated.

One way to visualise the operation of a CCD camera being exposed and then read out is to imagine a farmer trying to determine the way rain is falling onto an open field using a large number of buckets and a team of labourers. First, our hypothetical farmer marks out a grid of lines across the field to make a number of squares. Next, the squares are identified with letters and numbers, so that along one edge of the field the rows of squares have letters (A,B,C and so on) and up the edge at right angles they have numbers. Each square can be identified by a pair of letters and numbers such as A1, A2, B1, B2 and so on. The farmer then sends a labourer with a covered bucket to stand in each square and positions a measuring device, a scale or a measuring jug, at corner A1. Each marked square in the field corresponds to a CCD pixel and the grid lines are usually referred to as rows and columns. To start the measurement, analogous to opening the shutter in a CCD camera, the lid of each bucket is opened. Raindrops (representing photons, or rather the electrons generated in the silicon as the photons arrive) are then collected for some fixed time interval. To stop the exposure our farmer blows a whistle and all the buckets are covered over again. All that remains is to measure the amount of water (photoelectrons) in each bucket and to plot that on a chart representing the grid of squares across the field.

It would be complete chaos if all the labourers tried to run over to the farmer with their buckets, queue up to be measured and then run back to their squares ready to repeat the measurement. So the farmer has them all stand still and just measures the water in bucket A1. The farmer notes the amount on a paper and pours the water away. Since bucket A1 is now empty, the labourer in square A2 pours the contents of bucket A2 into bucket A1 for measurement and recording. While bucket A1 is being measured again, labourer A3 fills up the now empty bucket A2 and waits to receive the contents of bucket A4. This process continues until all the water in column A has been passed to the end for measurement and all the buckets in this column are empty. At this point, everyone in column B empties their bucket into the corresponding bucket in column A and the process of measuring bucket A1 (which now contains the water originally in bucket B1), and passing the water from bucket to bucket down column A for measurement is repeated. While this second set of measuring and recording is going on, all the buckets from column C are poured into those of the now empty column B, the column C buckets are re-filled from column D and so on across the field. Once all the column A buckets have been measured for a second time and are all empty again, they can be refilled once more from column B. The process of transferring water across the field, and then down one edge for measurement can be repeated until every bucket is empty. Since the bucketfuls reach the corner in a well-defined sequence (all the As, then all the Bs and so on) the farmer can reconstruct the pattern of rainfall right across the field without ever moving from one corner. This process of pouring water from bucket to bucket across the field and down the edge is the equivalent of reading out the CCD by varying the electrical voltages across it to force the electrons to move about and then counting them as they leave the chip.

Jewitt and Luu's first attempt with the new technology was made at the relatively small 1.3 m telescope at the McGraw–Hill Observatory at Kitt Peak, Arizona. The telescope was owned jointly by MIT and two other universities and was equipped with a CCD camera developed by George Ricker at Jewitt's home institution of MIT. The camera was known by the name MASCOT, the MIT Astronomical Spectrometer Camera for Optical Telescopes. The MASCOT CCD was an array of 390 by 584 pixels, laughably small by today's standards, and for a variety of reasons, not even all of this area could be used. Instead, the useful area was restricted to a region only 242 by 276 pixels, giving a field of view of just 9 by 10 arcminutes on the sky. This was about one thousandth of the area covered by a single plate from one of the Schmidt telescopes used in the other part of the search programme.

The observing strategy which Jewitt and Luu adopted was to take four consecutive exposures of each region of sky on one night, then make two more of the same field the following night to reveal any very slow-moving objects. The exposure time for each image was about 20 minutes. If the exposures were any longer then even a slow-moving object would begin to smear out across the image as it drifted across the sky. This smearing would mean that the light from the target would start to fall into adjacent pixels on the CCD and the signal to noise ratio would no longer improve as the exposure continued. In fact, the situation would soon start to get worse since the signal in the first pixel would no longer be improving while the continuing arrival of unwanted background photons from the sky would steadily increase the noise in it. Once the trailing effect becomes important, doubling the exposure time just doubles the number of pixels into which the light from the target is smeared and the signal in each pixel no longer increases.

Over five nights in April 1987 Jewitt and Luu observed 14 fields. Since the computers on Kitt Peak itself were not then powerful enough to process the images immediately, they took the data to Cherry Street to examine the images there. This enabled them to take advantage of the fact that not only are CCDs more efficient at detecting light than photographic plates, but the fact that they are read out electronically means that their images come out directly in digital form. The digital data which describe the images can be either displayed immediately or further processed to reveal subtle structures and objects that are not easily visible to the eye. One of the critical steps in this process is flat fielding, which cancels out variations in the sensitivity of different areas across the chip. Flat fielding is usually done using an image of a uniformly illuminated source, a white screen or the inside of the telescope dome for example. The resulting flat-field image maps out the pixel-to-pixel variations across the chip and its inverse can be applied to all the subsequent science images to 'flatten' them. The removal of false detections caused by electronic noise in the measuring circuits and by cosmic rays passing through the chip and liberating extra unwanted electrons is also usually necessary. Once these steps have been taken, it is possible to redisplay the image with different degrees of contrast or to try to improve the ratio of signal to noise in the region of interest by adding together groups of adjacent pixels. This latter process is called rebinning, but of course, rebinning does not come free. When more pixels are added together the fine detail in the resulting image is lost and the picture becomes grainy and harder to interpret.

The switch to CCDs also highlights another advantage of electronic detectors over photographic plates. Given adequate computing power, the images produced from an electronic detector can be searched in a number of different ways. The simplest of these is to subtract one image from another electronically. All things being equal the fixed objects will vanish and anything which has moved between the exposures will remain, appearing as a positive image from the first frame and a negative one from the second. Of course, in real life things are never that simple so complete subtraction of the fixed objects is seldom possible. Small changes in atmospheric turbulence, or seeing as it is known in the trade, and errors in the motion of the telescope as it turns to counteract the Earth's rotation, cause the images of stars to smear out slightly. This means that the image quality is often different from exposure to exposure. The images are also affected differently by cosmic rays, with about 200 such hits being recorded in each MASCOT frame. Despite these difficulties, Jewitt and Luu found it fairly easy to identify the relatively fast-moving main belt asteroids in the subtracted images, but they soon found that subtracting images was not the best way to search for faint, slow-moving objects.

Their preferred method was to blink images in much the same way as was done using photographic plates. However, instead of having to develop the plates, load and then carefully align them in a blink comparator before starting to search, modern image processing software can do quite a lot to ease the strain of the blinking process. The software can automatically match the overall brightness of pairs of images and even align them before the blinking starts. Then, with a couple of presses of a button, the images can be flashed alternately onto a screen at whichever frequency the person doing the searching feels most comfortable. This is usually about once per second. Not only this but three, four or even more images can be blinked and the image size and contrast can be varied quickly and easily to investigate potentially interesting objects. It is here that the advantage of taking multiple images of the same field comes into play. Any flashing source which does not appear in all the frames, and with a constant direction and rate of motion, can be ignored. Real moving objects exhibit a constant back and fro flashing motion which once seen and recognised is very distinctive.

This is not to say that blinking images by computer is either easy or quick. Any exposure deep enough to detect a new member of the distant solar system will certainly be long enough to detect hundreds of stars in our own galaxy and numerous faint galaxies beyond it. There is also a good chance that the field will contain a few main belt asteroids as well. Despite the advantages of new technology, blinking images is a very tedious business which demands a great deal of concentration. Typical CCD images contain hundreds of objects and to pick out a faint one jumping backwards and forwards by a few millimetres on a computer screen is a tricky task, especially when tiredness and boredom are setting in. Sometimes, just as when searching photographic plates, changes in the seeing during the time between the two images causes the faintest objects to vanish from one frame, but not the other. If this happens, then chance alignments of faint sources mimic the back and fro flashing of a moving object and these can only be eliminated with careful examination of each image. The key to success is having considerable patience and training the eye–brain combination to recognise what looks right whilst ignoring all the rest. Some people are better at this than others, so quite often the best observing teams comprise pairs of people, one of whom operates the telescope and camera while the other blinks the images relentlessly to see if anything can be found.

Although Jewitt and Luu both blinked their MASCOT images to check that nothing had been missed, they did not find any new distant asteroids. This was perhaps not very surprising; the relatively small size of the MASCOT CCD meant that only tiny areas of sky could be searched at a time. Indeed, the area covered by their 70 CCD frames covered only one third of a square degree, or less than one hundred thousandth of the whole sky. However, the onward march of technology soon led to an increase in size of a typical CCD detector and the amount of sky that could be imaged in a single image grew steadily, offering more and more chances of success. Dave Jewitt was not deterred by his early, negative results.

In 1988 Dave Jewitt made another move. He left MIT to take a faculty position at the Institute for Astronomy, or IfA, in Honolulu, on the island of Oahu, Hawaii. The IfA owes its existence to the fact that the summit of Mauna Kea, a huge dormant volcano on the nearby and less cosmopolitan Big Island of Hawaii, is an excellent place to build large telescopes. The Hawaiian island chain comprises several large, and many small, islands stretching across a swath of the Pacific Ocean. They were created as one of the Earth's tectonic plates drifted over a hot spot in the molten mantle beneath. Magma escaping through the hot spot built a series of volcanoes, several of which rose well above sea level. The extinct volcano of Haleakala rises some 3000 m over the island of Maui and has long been the site of a solar observatory and a research facility of the US Department of Defense. The story goes that astronomers working on Haleakala, including ironically Gerard Kuiper, would glance across at the even higher mountain of Mauna Kea on the Big Island and ask themselves, 'Why don't we build an observatory there?'. In 1964 a road, or more precisely a dirt trail accessible by four-wheel-drive vehicles such as jeeps, was cleared to the summit of Mauna Kea and testing of the mountain as a possible place to put large telescopes commenced. The testing soon showed that Mauna Kea was indeed an exceptionally fine observing site. Kuiper described it as, 'Probably the best site in the world from which to study the moon, the planets and stars'. Three groups, Kuiper from Arizona, a team from Harvard University and another from the University of Hawaii all applied to NASA for funding to build a telescope on Mauna Kea. In 1965 NASA awarded the University of Hawaii a grant to develop a 2.24 m telescope on the summit. Soon after, the state of Hawaii decided that the University of Hawaii should be placed in charge of the development of Mauna Kea and so the IfA was created near the Manoa Campus of the University, in Honolulu.

During the first few years of the development of Mauna Kea, four large telescopes were built there, including the 2.24 m optical telescope which belonged entirely to the University of Hawaii and a larger optical telescope, the 3.6 m Canada–France–Hawaii Telescope, or CFHT. Joining these two were a NASA funded infrared telescope devoted to observational support of space missions and the 3.8 m United Kingdom Infrared Telescope (UKIRT). As the value of Mauna Kea as an observatory site began to be recognised, more and more observatories were established on the mountain and it is now one of the most important astronomical observatories in the world. As 'landlord' of Mauna Kea, the IfA collects a tithe of observing time from all of the observatories on the mountain, so in addition to their own telescopes, they have guaranteed access to 10–15% of the time on all of the other facilities. With so much telescope-time available to a relatively small number of astronomers, the IfA's telescope-time allocation committee looked favourably on Dave Jewitt's proposals. Soon he began to receive allocations of observing time on the 2.24 m telescope for various solar system projects. At first, his programmes were not primarily searches for distant objects, but as the capabilities of the telescopes were improved, particularly in respect of bigger and better CCD cameras, Jewitt began to think about doing some more surveys.

However, by then Dave Jewitt was not the only one searching for trans-Neptunian objects. Between November 1988 and March 1989 Harold Levison, known to his friends as Hal, and dynamicist Martin Duncan surveyed almost 5 square degrees of sky using the 1 m tele-

Figure 4.2 The University of Hawaii's 2.24 m (88 inch) telescope used by Jewitt and Luu during their search for the Kuiper Belt. One of the four-wheel-drive vehicles used to drive up to the summit is parked outside. (John Davies.)

scope at the US Naval Observatory at Flagstaff, Arizona. The 2048 square pixel CCD they used covered a square of 26 arcminutes a side and with it they observed a total of 26 fields covering a total of 4.88 square degrees. They imaged each field for 40 minutes at three separate times spread over a couple of days. They did not search their images by blinking them, but instead used computer software to identify all the objects in each image and then tried to link together sources which did not appear at the same place in successive images. This linking was done on the basis of pairs of sources having roughly similar brightness and whose separation was consistent with motion at the typical rate of a distant solar system object. The software did indeed find many such linked objects, between five and 70 in each field, all of which had to be examined individually. Every one of them turned out to be false alarms, attributable to chance alignments of cosmic ray hits, electronic noise in the CCD readout and other defects.

So, despite their efforts, no distant objects were found, although Duncan jokingly relates they nearly named a candidate after the mythological character Ups, as in 'Whoops, what's that?' Based on the negative results of the search, Levison and Duncan attempted to place limits on the number of Chiron-like objects and comets between about 25 and 42 AU from the Sun. For Chiron they compared the percentage of the sky they surveyed (0.01%), the number of objects they found (none) and the number of Chiron-type objects known at the time (one). Based on the statistics of these very small numbers, they concluded there were less than 7000 Chiron-sized objects in the outer solar system. They also noted that this value, and their estimates of the populations of still smaller objects, were very dependent on a number of their assumptions and could not be taken 'too seriously'.

Another search was carried out at the 2.7 m telescope of the McDonald Observatory in Texas by Anita and Bill Cochran. Anita came to the University of Texas as a graduate student in the late 1970s and while there she met her future husband. Astronomer Bill Cochran was a postdoctoral research worker in the university. Between November 1990 and March 1993 they, mostly Anita, spent a total of 22 nights, including the Christmas eve of 1992, observing. Images were taken with an 800 by 800 element CCD and were blinked to search for moving objects. However, the search was dogged by poor weather and poor seeing (which smears out faint objects over many CCD pixels and makes them harder to detect above the background of the night sky) and nothing was found. Realising that she could not compete with telescopes at other sites, where better seeing made detecting faint objects easier, and running out of time due to pressure of other projects, Anita Cochran put her search programme aside for the moment and moved on to other things. One of these was to be a project to search with a telescope where seeing was not going to be a problem.

One other search project carried out around this time is worthy of mention, even though it was initially carried out using images taken for a quite different reason. In September 1990, Piet Hut of the Institute for Advanced Studies at Princeton, New Jersey, was listening to a talk about a proposed telescope which would survey a large region of sky.[†]

[†] This became the Sloan Digital Sky Survey.

The primary purpose of the survey was to make a huge census of galaxies, but Hut suggested that it might also turn up some rather closer things. After the talk, Hut was approached by Tony Tyson from the nearby Bell Laboratories, who said that it was not necessary to wait for the survey to be done. He and his colleagues Raja Guhathakurta and Gary Bernstein already had some deep CCD images that might be suitable for just such a search. Hut talked with Guhathakurta about ways in which their data could be used and they developed some software to add together individual images in moving reference frames. Adding images in this way smears out the fixed stars and galaxies, but adds together the signal from moving objects. Of course, this procedure only works if you know in advance how fast the target is moving. Luckily, since the motion of objects beyond Neptune is dominated by the reflex motion of the Earth overtaking them, a good guess at the correct rate of motion is possible even without knowing exactly how far away the targets are. The search algorithms were then tested on frames which had been modified to include a fake moving object. The software seemed to work well, so it was applied to some real images to see if anything could be found.

The method they settled upon was to use a number of their images taken with relatively short integration times and combine them to make a deep image in the reference frame of the fixed stars. Then they electronically subtracted these fixed objects, and removed any defects and cosmic ray hits from each of the original exposures. This left a series of 'blank' residual images. These were then added together with successive offsets to allow for the motion of any hypothetical faint objects travelling across the image in a certain direction with a given speed. This rate of movement across the image is called the object's apparent motion vector. The shifting and adding was then repeated for a range of slightly different apparent motion vectors to maximise the chance of finding faint moving objects.

Although this method will work for any part of the sky, the best chance of success will be for regions close to the ecliptic, where the density of potential targets is the highest. Tyson, Hut and colleagues tried their method on some images of a field 3.5 degrees from the ecliptic which they had taken in 1991. The field was observed nine times over a two-night period and covered a total area of 40 arcminutes a side. They added the images to cover a range of possible motions from 1 to 4 arcsec per hour. Nothing was found, even though they would have expected to see objects as faint as 25th magnitude. They were unlucky, the method was a good one and the sensitivity they reached was high enough to have detected something if it had been there. It just so happened that the field they picked was empty on the nights they chose to look there. They got, in the words of Piet Hut, 'Close, but no cigar'.

While Hal Levison, Anita Cochran and the others were trying and failing, Dave Jewitt, who had been joined in Hawaii by Jane Luu in the autumn of 1988, continued to search using the University of Hawaii's (UH) 2.24 m telescope on Mauna Kea. Since it was from here that success would eventually come, it is worth considering for a moment exactly what they were up against. The summit of Mauna Kea is not an easy place to work. At an altitude of 14 000 ft (4205 m), almost half way to the height jet airliners fly, the air is thin, with only 60% of the oxygen found at sea level. It is often cold, and in the winter snow sometimes blankets the summit for days at a time. Almost nothing lives at these high elevations. There is no vegetation on the reddish-brown cinder cones and only a few very specialised insects survive by eating other less hardy insects blown up the mountain to die. The thin and often very dry air produces a number of interesting physiological effects on the humans who choose to go there in search of clear skies. These include headaches, itching eyes and nosebleeds brought on by increased blood pressure as the heart tries to pump blood around the body fast enough to ensure an adequate supply of oxygen to the brain. There is also a little talked about tendency to run to the toilet rather frequently as the kidneys try to thicken the blood by ejecting excess water. To minimise the risk of altitude sickness, which is at best unpleasant and at worst can be fatal, astronomers working on Mauna Kea do not sleep at the summit. Instead they stay at a dormitory-cum-cafeteria called Hale Pohaku (from the Hawaiian for 'house of stone') situated at the 3000 m level, more than a kilometre below the telescopes themselves. As an aid to acclimatisation, the astronomers usually arrive at Hale Pohaku the night before their observing run. This gives their bodies a chance to begin the adjustment process before they go to work at the summit the next evening. Despite these precautions, most regular users of Mauna Kea admit that their thought processes at the summit are not as clear as they would like. If pressed, most of them will admit to at least a few hours of almost moronic thinking trying to solve a technical or mathematical problem they could sort out in a matter of minutes in the thicker air at sea level. It was in this environment that Dave Jewitt and Jane Luu continued their quest for the edge of the solar system.

Their technique changed little over the years. They tried to observe on moonless nights, when the sky would be at its darkest and so faint objects would not be lost in the glare of scattered moonlight. They always observed in the spring and autumn. These periods were chosen since it is the time of year when the ecliptic plane, their best hunting ground, was well separated from the Milky Way. Both the ecliptic and the Milky Way, the plane of our Galaxy, trace out great circles across the sky, but these two circles are not the same. This is because the solar system is tipped relative to the plane of the Galaxy. The great circles cross in two regions, in the northern sky at the border of the constellations Gemini and Taurus, quite close to the famous Crab Nebula, and in the south in the Sagittarius–Ophiuchus region. Since the faint band of light which we call the Milky Way is actually the combined starlight of most of the stars in our own Galaxy, it is a very crowded region of sky, already bedecked with a myriad of faint stars. It is hardly the place to look for a faint point of light moving against a background of many other points of light. Far better to look away from the Milky Way, in the constellations of spring and autumn. Here the sky is less crowded and any interlopers are easier to spot and less likely to be masked by the light from background stars.

For Jewitt and Luu, each night began with an early dinner at the halfway house of Hale Pohaku and a chat with some of the other astronomers or perhaps some last-minute preparations for the observing. After dinner came the 20 minute drive, by four-wheel-drive vehicle, up the twisting and bumpy dirt road that connects Hale Pohaku to the summit. The UH 2.24 m telescope is on the third floor of a traditional white observatory building, cylindrical in shape and topped by a dome. From certain angles, the outline of the dome is broken by an unusual hammerhead extension which is the parking spot for a crane used to move heavy equipment about inside. The building has laboratory space and workshops on the lower floors. The control room is reached by taking a lift part way, passing through a small workshop which smells of oil and machine tools, and then climbing a short flight of spiral stairs.

Described by one of the staff as 'the highest office in the world' the control room itself is banana shaped, long, narrow and curving to match the inside wall of the cylindrical building. At one end, a door opens onto a catwalk around the outside of the building. From the catwalk all the other observatories and a superb view of the Mauna Kea sunset are visible. Ranged along the inner edge of the control room are consoles for the telescope operator, who is responsible for pointing the telescope and generally keeping things running smoothly. Next to them are the computers of the observer's station, from where the scientific instruments are controlled. At the other end of the room, an electric kettle and jar of instant coffee hint at a long observing night. An internal door gives access to the observing floor and to the telescope itself.

The telescope is a stumpy tube, rather like a barrel, held in the arms of a massive metal fork which towers over the observing floor. The telescope is painted orange-brown; it is said that the paint was blended especially to match the colour of the dust and rocks of the summit in order to bring good luck to the observatory. The base of the telescope, where the scientific instrument for the nights' observing is fitted, is about four metres above the floor. It can be reached by a hydraulic platform which can be raised to provide a working area around the telescope. The CCD camera operates best when it is cold, so before observing can begin it must be topped up with liquid nitrogen coolant.

Once everything was ready and the dome had been opened, Jewitt, Luu and their telescope operator settled down for the night's observing. Night after night they took a series of images, each exposed for 15 minutes, and returned to the same field three times over the following two hours. They worked as a team. One of them concentrated on the image processing and blinking, the other controlled the camera and kept track of the observing. As the images came in, they were displayed on a computer dedicated to data reduction located in the telescope control room. Jewitt and Luu blinked as they went along although not everything could be done at once, despite having what for the time were quite powerful computers. Blinking the images demanded so much from the computers that it could only be done during the exposures when the CCD was inactive. Once the process of reading out the CCD was about to begin, the blinking had to be stopped. If it was not, the overloaded computing systems found it impossible to cope and crashed.

The initial observing runs did not result in a rapid success. The pair detected hundreds of main belt asteroids, but they ignored these in order to concentrate on their more distant goal. However, by the early 1990s CCD technology was moving steadily forward and larger and larger chips were becoming available almost every year. The 2.24 m telescope soon moved from a 385 by 576 CCD to one 800 pixels square. This seemed huge at the time, but was in fact only a step to still better things. Soon, the CCD chips had increased in size to 1024 by 1024 pixel arrays. This steady increase in detector size meant that it was easy for Jewitt to convince himself that success was only a matter of time and to maintain his enthusiasm for the search. Jewitt helped persuade Gerry Luppino, who had been a graduate student at the Massachusetts Institute of Technology and who was known as an excellent builder of CCD cameras, to join the IfA. By 1992 the 2.24 m telescope was equipped with a camera containing a chip with 2048 by 2048 pixels able to cover an area of sky about 7.5 arcminutes square. With it Jewitt and Luu continued to search for their elusive goal. Sometimes they worked quietly, sometimes the control room pounded to the sound of Jewitt's CD collection, very heavy metal music from groups with names like Cannibal Corpse, Mortician and Napalm Death. Jane Luu was not a big fan of Jewitt's musical tastes. When it got too much for either her or the telescope operator, trapped in the same room night after night, they switched to something from Luu's more restrained classical music collection. Sometimes they forsook the CD player for Mauna Kea gossip and long rambling quasi-philo-sophical conversations about nothing in particular.

The Jewitt and Luu combination was something of an attraction of opposites. From the outside it seemed incredible that the outspoken and flamboyant Jewitt and the reserved and cultured Luu would be compatible, but they got along well and were a very good observing team. They were both determined to get the most of every minute of telescope time. According to one of the telescope operators who worked with him, Jewitt would even plan his observing to minimise the number of times the telescope's protective dome had to be moved. Every move was a couple of precious minutes lost. The pair made a very productive partnership, publishing papers on the physical prop-erties of cometary nuclei, on the statistics of the Earth-approaching asteroid population and on the activity of the Centaur Chiron. Jane Luu wrote a thesis entitled 'An observational investigation of the comet-asteroid connection', and left the IfA for a postdoctoral position at Harvard University in 1990. She came back to Hawaii from time to time to continue the search.

The long-awaited breakthrough came around midnight on 30th August 1992. It was the second night of a five-night observing run. Jewitt and Luu were taking the third of a sequence of four images of a field in the constellation of Pisces, which they called simply ECL 123. Dave Jewitt was blinking the first two images from the set. Suddenly he said, 'Jane, come and have a look at this'. There was something starlike moving slightly between frames as they flashed alternately onto the screen. This seemed to be exactly what they were looking for, or was it? With only two images it was a bit early to be sure. Perhaps it was a false alarm caused by cosmic ray hits or some other defect. However, they were excited enough to measure the rate of motion, which turned out to be 2.3 arcseconds per hour. This would place the object about 60 AU from the Sun. When the third and then the fourth image confirmed that they had indeed found a faint slow-moving object, they were dumbstruck. Jewitt later described it as a sort of mental collision between seeing the object, which looked just right, and the experience of years of fruitless searching. Jane Luu recalls 'I think we jumped up and down for a little while'. The only question was whether it was indeed very distant or something much closer. There was a chance that it might be a small, near-Earth asteroid moving almost parallel to the Earth and which happened to appear almost stationary on the sky at that particular moment.

The only way to check if they had finally found what they wanted was to make more observations. Anxiously they re-observed the object several times over the rest of the night. They were looking for any change in its rate of motion. Such a change would reveal that the object was so close that as the Earth turned the different relative positions of the object and telescope would alter the observing geometry. No such changes were found. Before breakfast they called Brian Marsden at the Minor Planet Center to let him know they had found something that looked interesting, but they agreed to keep it quiet while they collected more observations. Over the remaining three nights they re-observed the object repeatedly. They checked their measurements again and again looking for anything which they might have overlooked and which might destroy their conclusion that the object was far beyond Neptune. They measured its brightness, position and motion as accurately as possible and then they sent their data to the Minor Planet Center. On 14th September Marsden announced the discovery via IAU circular number 5611. He gave the new 'asteroid' the prosaic designation 1992 QB$_1$.[†]

[†] The subscript 1 indicates that in that fortnight more than 24 objects had been reported and the Minor Planet Center had had to use the letter B more than once. In the following years, a dramatic increase in the number of asteroid sightings from automatic telescopes would mean that much larger subscripts would be needed.

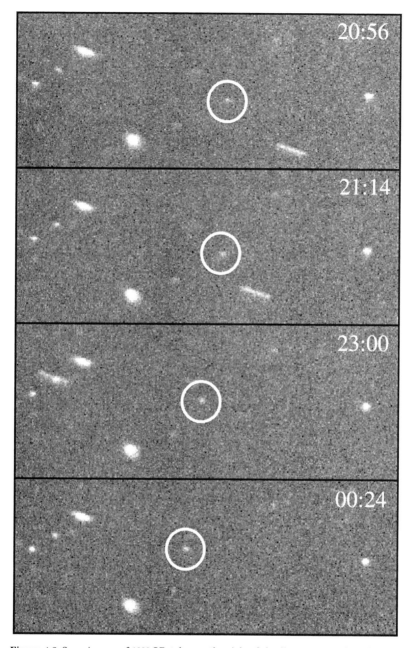

Figure 4.3 Some images of 1992 QB$_1$ taken on the night of the discovery. A main belt asteroid appears in the top three images, moving from right to left. The motion of 1992 QB$_1$ is very much slower. (Dave Jewitt.)

From the limited number of positions available he could not confirm exactly where the new object was; in fact, Marsden could not even be sure which way it was going around the Sun. The first positions could be fitted with an orbit at about 41 AU going in the same direction as the rest of the planets or with one some 15 AU further out and going the other way. However, he did note that some solutions were 'compatible with membership of the supposed Kuiper Belt', although he qualified this remark by saying that the object could also be a comet in a near-parabolic orbit. Further observations, over a longer arc, would be required to solve this problem.

Luckily, these further observations were not long coming. As the Moon waned and the skies became dark again, 1992 QB_1 was observed from the Anglo-Australian Observatory at Siding Springs, New South Wales, on 21st September. It was seen again from the European Southern Observatory in Chile a few days later. Using these positions, Marsden calculated another orbit which placed 1992 QB_1 a few AU further out, but which was otherwise quite similar to his first attempt. However, the issue of the nature of the new object was still not closed. Its positions could also be fitted by other more-eccentric orbits. Marsden thought it possible that 1992 QB_1 was a Centaur like Pholus and Chiron, but that it had been discovered at the farthest point in its orbit, rather than when close to the Sun. If this was true then there was a chance that a pre-discovery image might appear on an old photograph taken when 1992 QB_1 was closer to the Sun and so brighter. Marsden, who enjoys entering into lighhearted wagers, found what looked like a suitable candidate from 1930 and he made a bet with Dave Jewitt that observations over the next few months would reveal that the object was a Centaur.

Not just the orbit of 1992 QB_1 was uncertain, almost nothing was known about its physical properties either. From the limited observations available, Jewitt and Luu deduced that their discovery was quite red. This was unusual for an asteroid and hinted that 1992 QB_1 might resemble the Centaur Pholus, discovered earlier the same year. They searched for, but could not find, any trace of gas and dust, which more-or-less ruled out it being a large and very distant active comet. Even the size of the new object was uncertain; without knowing details of how reflective its surface was, the discoverers could only make an educated guess at its diameter. Assuming that the reflectivity was about 4%, a typical number for a dark asteroid, they estimated that the new body was quite small, about 250 km in diameter. This would make it one ninth the size of the planet Pluto.

SEP 2 1 1992   Circular No. 5611

Central Bureau for Astronomical Telegrams
INTERNATIONAL ASTRONOMICAL UNION

Postal Address: Central Bureau for Astronomical Telegrams
Smithsonian Astrophysical Observatory, Cambridge, MA 02138, U.S.A.
Telephone 617-495-7244/7440/7444 (for emergency use only)
TWX 710-320-6842 ASTROGRAM CAM   EASYLINK 62794505
MARSDEN@CFA or GREEN@CFA (.SPAN, .BITNET or .HARVARD.EDU)

1992 QB₁

D. Jewitt, University of Hawaii; and J. Luu, University of California at Berkeley, report the discovery of a very faint object with very slow
(3″/hour) retrograde near-opposition motion, detected in CCD images obtained with the University of Hawaii's 2.2-m telescope at Mauna Kea. The
object appears stellar in 0″.8 seeing, with an apparent Mould magnitude
$R = 22.8 \pm 0.2$ measured in a 1″.5-radius aperture and a broadband color
index $V - R = +0.7 \pm 0.2$.

1992	UT	α_{2000}	δ_{2000}
Aug.	30.45568	$0^h 01^m 12^s.79$	$+0° 08′ 50″.7$
	30.59817	0 01 12.19	+0 08 46.9
	31.52047	0 01 08.37	+0 08 22.7
	31.61982	0 01 07.95	+0 08 19.9
Sept.	1.35448	0 01 04.90	+0 08 00.6
	1.62225	0 01 03.76	+0 07 53.3

Computations by the undersigned indicate that 1992 QB₁ is currently
between 37 and 59 AU from the earth but that the orbit (except for the
nodal longitude) is *completely indeterminate*. Some solutions are compatible with membership in the supposed "Kuiper Belt", but the object could
also be a comet in a near-parabolic orbit. The particular solution below is
the direct circle (but a retrograde circle some 15 AU larger in radius also
fits); Jewitt and Luu note that a cometlike albedo of 4 percent then implies
a diameter of 200 km and that the red color suggests a surface composition rich in organics. Further precise astrometry during the late-September
dark run should eliminate some possibilities, but a satisfactory definition
of the orbit will clearly require follow-up through the end of the year. The
object's phase angle reaches a minimum of $< 0°.01$ around Sept. 22.5 UT.

$$\text{Epoch} = 1992 \text{ Aug. } 26.0 \text{ TT} \qquad u = 0°.335$$
$$\Omega = 359.440 \left.\right\} 2000.0$$
$$a = 41.197 \text{ AU} \qquad i = 2.334$$

1992TT	α_{2000}	δ_{2000}	Δ	r	ϵ	β	V
Sept. 15	$0^h 00^m.09$	$+0°01′.7$	40.200	41.197	172°.5	0°.2	23.4
25	23 59.33	− 0 03.1	40.195	41.197	177.5	0.1	23.4
Oct. 5	23 58.58	− 0 07.9	40.220	41.197	167.5	0.3	23.5
15	23 57.87	− 0 12.5	40.275	41.197	157.4	0.5	23.5

1992 September 14   *Brian G. Marsden*

Figure 4.4 The IAU circular announcing the discovery of 1992 QB₁. The circular contains
some brief discovery details, a list of the positions reported by Jewitt and Luu and some
comments from Brian Marsden. These are followed by approximate orbital elements and
predicted positions for the next few weeks based on these elements. (CBAT/Brian
Marsden.)

Unless 1992 QB_1 was indeed a Centaur then its faintness meant that there was little hope that images of it would be found on older photographic plates. It was clear that it would take some time to define the object's orbit with any great precision. However, six month's later, Jewitt and Luu discovered a second object, designated 1993 FW, which was of similar brightness and which was moving in the same general way. When this was announced there was little doubt that the existence of a trans-Neptunian population had finally been confirmed. Marsden, being nothing if not a man of his word, paid up on his bet with Dave Jewitt at a scientific conference in Flagstaff, Arizona, in the summer of 1994. By then, four more distant asteroids had been found and Marsden was giving a review of the discoveries and of the likely orbits of the the new objects. Almost at the conclusion of his talk he called on Dave Jewitt and solemnly handed him five banknotes in front of the assembled audience. It was all done in such good humour that few of the people in the theatre realised how much money was changing hands. Each of the five banknotes was a $100 bill![†]

[†] Brian Marsden got some of his money back two months later when, at a different meeting, Dave Jewitt bought him a very expensive lunch.

Deeper
and deeper

With the discovery of 1992 QB$_1$ and 1993 FW, Dave Jewitt and Jane Luu had opened the door on a new field of solar system research. Not since Piazzi discovered the first main belt asteroid in 1801 had an entirely new class of solar system object been discovered. In contemplating the implications of the discovery, Dave Jewitt mused that, 'Discovering the Kuiper Belt is like waking up one morning and finding that your house is ten times as big as you had thought it was'. However, finding two objects was a beginning, not an end. To find out just how large the Kuiper Belt really was would require the discovery of many more trans-Neptunian objects. Jewitt did not rest on his laurels; indeed, his success seemed to spur him on to greater efforts. Together with Jane Luu, who was still working in California, he continued searching. During an observing run in September 1993 they found another two faint, slow-moving objects. These were soon designated 1993 RO and 1993 RP. Brian Marsden announced the discoveries via IAU Circulars 5865 and 5867 on the 18th and 20th of September, respectively. He noted that his initial calculations suggested that the new objects appeared to be closer to the Sun than the first two discoveries, but that with so few observations, the orbits could not be established with any certainty. Marsden noted that the two objects could be in circular orbits about 39 AU from the Sun, more eccentric orbits at various distances or even in a variety of retrograde orbits. The observations could also be fitted by parabolic trajectories as if the objects were long-period comets entering the solar system from the Oort Cloud. Marsden's dilemma was compounded a few days later when another group entered the fray, reporting the discovery of two more faint, slow-moving objects.

One of the members of this group, who would go on to make other contributions to this new field, was a young scientist named Alan Fitzsimmons. Like many other astronomers, Fitzsimmons had become interested in the heavens as a child and his hobby gradually became his career. As he recalls it, he was about twelve or thirteen when he visited a friend who was aching to show off the telescope he had just received as a present. Fitzsimmons was talked into having a look at the Moon and, impressed despite himself, he decided to save enough money to buy a telescope of his own. Like Dave Jewitt and lots of other young people, he slowly became hooked on astronomy. At the time Fitzsimmons' career goals did not include science; he saw his future in business and he planned to study physics and computing at university with this in mind. It was only a chance conversation with the brother of a friend which alerted him to the possibility of studying astronomy at university. After completing a first degree at Sussex, he went on to do a PhD at Leicester University. His project involved studies of Halley's comet and brought him into contact with some of the UK's few solar system astronomers.

Leaving Leicester, and the solar system community, he moved to Queens University in Belfast, Northern Ireland. Here he began to do research on the chemical composition of stars and eventually accepted a permanent post in the Physics department. While he was at Belfast, he saw the IAU circular announcing the discovery of 1992 QB_1 and thought to himself 'We can do this'. After talking with Professor Iwan Williams from Queen Mary and Westfield College in London, an observing crony from his Halley's comet days, he contacted Dave Jewitt. Together they applied for telescope time to make a further search. Williams and Fitzsimmons were successful at the first attempt. They were awarded a week of observing time on the 2.5 m Isaac Newton Telescope (INT) on La Palma, in the Canary Islands.

The observing run took place in late September 1993 using a 2048 pixel square CCD camera. Dave Jewitt was unavailable, he was observing in Hawaii, so Fitzsimmons and Williams were joined on the run by Belfast student Donal O'Ceallaigh. They used the same observing technique as Jewitt and Luu, taking and then blinking 30 minute exposures aimed at a swath of positions along the ecliptic plane. The project got off to a good start with a discovery on the second night of the run. The object was faint and it required the taking of a few more images to confirm its reality. Once these extra frames had been taken there was no doubt that there was something there. The object's position was measured and transmitted to the Minor Planet Center, which designated it 1993 SB. The following night they found another object. The second discovery, 1993 SC, was so much brighter than the first one that Fitzsimmons later described it as being, 'blindingly obvious' in the blinked frames.

Things were going so well that they were not very surprised to find another bright candidate two days later. Amazed by their good fortune, and thinking that this business was much easier than they had expected, they carefully measured the positions of the new object and dutifully reported them to the Minor Planet Center. Brian Marsden replied by remarking that, based on the positions, their second and third objects seemed to be one and the same. It only took a minute to confirm this and realise that, carried away by their enthusiasm, they had made a mildly embarrassing mistake. Somehow they had made an error when calculating the location of the ecliptic plane on the sky. Instead of continuing on into unexplored territory, they had doubled back and accidentally rediscovered 1993 SC. The odd thing was that, due to its slow motion, the object had hardly moved from where it had been two nights earlier. In all the excitement not one

Figure 5.1 The Isaac Newton Telescope building on the island of La Palma in the Canary Islands. Unusually for a modern observatory, office and workshops are located around the telescope dome. The fifth and sixth trans-Neptunian objects were discovered from here by Iwan Williams and Alan Fitzsimmons. (John Davies.)

of the three observers had noticed that object number three was in a starfield that they had searched before and which, in hindsight, looked rather familiar. As with 1993 RO and 1993 RP, the initial observations suggested that the new objects were located in the general region of Neptune, but, once again, with only a few days of observations accurate orbits could not be determined.

Alan Fitzsimmons subsequently led several other observing runs to La Palma, discovering a number of other trans-Neptunian objects. Meanwhile, Jewitt and Luu continued to make discoveries from both Mauna Kea and from a small telescope in Chile. Other astronomers were not slow to realise that there was much to be done at the new frontier of the solar system and soon other groups joined in the hunt. Over the next few years the rate of discovery increased steadily as more and more people developed an interest in these obscure solar system objects. Paradoxically, it was not long before the problem was not so much discovering new objects, but keeping track of the ones that had already been found.

It was soon apparent that a proper understanding of the Kuiper Belt would require two lines of attack. On the one hand detailed physical studies were needed to probe the sizes, shapes and chemical compositions of individual objects. The first step along this road was for the orbits of individual objects to be determined accurately. Once this was done their positions could be predicted well enough to allow them to be studied with large telescopes. On the other hand, there was interest in a more general understanding of the population of the trans-Neptunian region as a whole. Accurate orbits for a large number of objects would eventually be needed to provide theoretical astronomers with information about the dynamics of the trans-Neptunian region. The theoreticians wanted real orbits which they could compare with the phantom objects predicted by their computer models.

However, since an object at a great distance from the Sun moves only slowly across the sky, observing its motion over a few days or weeks is not sufficient to determine its orbit precisely. Orbits can be described by numbers called orbital elements, and once these are known accurately the object's position can be projected well into the future. A typical set of orbital elements comprises six numbers, each usually quoted to five, six or even seven decimal places. It is the job of people like Brian Marsden and his colleagues at the Minor Planet Center to work them out. To do this, Marsden needed follow-up observations of each object a few times during the year of its discovery and again from time to time over the next few years. This long period of observation was demanded by the objects' great distances from the Sun. Even five years of observations would cover only 2% of a typical orbit. The problem was that with visual (V) magnitudes of only about 24, about sixty million times fainter than the faintest star visible to the naked eye, any kind of further observations required observing time on moderately large telescopes. Without this vital astrometric follow-up there was a real risk that some of these objects might be lost as uncertainties in their expected position built up to the point that it would be impossible ever to find them again. The situation was eerily reminiscent of that almost two centuries earlier. Then too, asteroids were sometimes discovered one year only to be lost again later. In the nineteenth century, when orbital calculations were done on paper by a small band of mathematicians, the problem was one of a lack of computing power. In the late twentieth century there was plenty of computing power; the problem was a lack of observing time on big telescopes.

Big telescopes around the world are the front line of modern astronomy and naturally large numbers of astronomers want to use them. Since there is never enough telescope time to go around (a typical national telescope receives three to four times more observing requests than it can accommodate) astronomers compete with each other through a process called peer review. What this rather grandiose title actually means is that once or twice a year astronomers write proposals asking for a certain amount of telescope time. These proposals, which are usually a few pages long, are then reviewed to decide which projects get an allocation of telescope time in the coming months. This competition is judged by other astronomers who try to be unbiased, but who generally have their own ideas on what is important for the advancement of science and what is not. There are trends, or fashions, in science just as there are in other things, and at any given time there will be areas of astronomy which seem to be producing the most interesting results. It is these fields which may seem to offer the best chance of critical discoveries that will produce a real scientific breakthrough, perhaps even a Nobel prize or two. Confirming the orbits of a few distant asteroids, however interesting they might be to a small number of planetary scientists, is not seen by many astronomers as being on the cutting edge of modern astrophysics. So, with the best will in the world, allocation committees were sometimes reluctant to award time for apparently routine requests for follow-up measurements of newly discovered Kuiper Belt objects. There may well have been a feeling that, 'Some-one else must be doing it', but in fact, a lot of the time, nobody was.

Marsden and a few other astronomers were acutely aware of this problem. As early as 22nd September 1993, when he was still wrestling with deciding exactly where in the solar system to put the four objects discovered that autumn, Marsden had written that 'There is rapidly developing a severe problem of securing adequate astrometric follow-up which is absolutely essential for any understanding of this excit-ing development in the outer system'. People like Jewitt, Luu, Fitzsimmons and their collaborators were conducting a determined rearguard action of astrometric follow-up, but they were often ham-pered by bad weather or baulky equipment. Soon it seemed that objects were being lost almost as quickly as they were discovered. Anita Cochran, who had been scooped in her own Kuiper Belt search, contributed to follow-up observations. Mark Kidger from Spain, David Tholen and Brett Gladman were amongst the few other dedicated souls willing to use the limited telescope time they were allocated for the essential, but routine astrometric follow-up. The magnitude, if the reader will forgive the expression, of this problem was brought home by the appearance of astrometric observations from a new observa-tory site unfamiliar to most professional astronomers. The little known Cloudcroft observatory was in fact the home of an amateur astronomer named Warren Offutt who was carrying out a programme of astrometric follow-up of trans-Neptunian objects for fun!

Even amongst a group of people as notoriously difficult to pigeon-hole as astronomers, Offutt stands out as something out of the ordi-nary. He was born in a suburb of New York City in 1928. This was two years before Pluto, let alone its brethren in the Kuiper Belt, had been discovered. He then had a successful career in industry before retiring in 1990. Unlike the professional astronomers exploring the Kuiper Belt, who are beholden to their employer or some other organisation to fund their research, Warren Offutt could afford to have his own observatory and then use it as he chose. Located near the village of Cloudcroft, New Mexico, not far from the professionally operated Apache Point observatory, Warren Offutt's 60 cm telescope sits on a knoll a few dozen metres from his house. It is inside a traditional masonry building capped by a dome which contains both the telescope and its control system. Offutt did not build this telescope himself.

Rather, as befits a man with decades of experience in the manufacturing industry, he drew up a detailed specification of what he wanted and negotiated with someone else to build it for him.

Although he had once started a project to photograph all the planetary nebulae visible from Cloudcroft, Offutt abandoned this idea half way through when he discovered that someone had already done it. Luckily, as is often the way in astronomy, a new project came along at just the right moment. At about 8 o'clock one morning in 1995 local astronomer Alan Hale called and asked him to verify the position of a new comet he had discovered. Using the positions, estimated direction and rate of motion supplied by Hale, Offutt looked for the supposed comet the next night. After a short time it was clear that the comet was not where he expected to find it. Where had it gone? Not to be beaten, Offutt scanned along an arc of sky covering the likely track and detected the comet on his third photograph, taking what may well have been the first post-discovery image of Comet Hale–Bopp. Although it was soon clear that the comet was beyond Jupiter, and so must be unusually bright, determining how close it was going to come to the Sun turned out to be problematical at first. Many observers took photographs to help find out just what was going on and Offutt

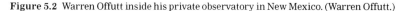

Figure 5.2 Warren Offutt inside his private observatory in New Mexico. (Warren Offutt.)

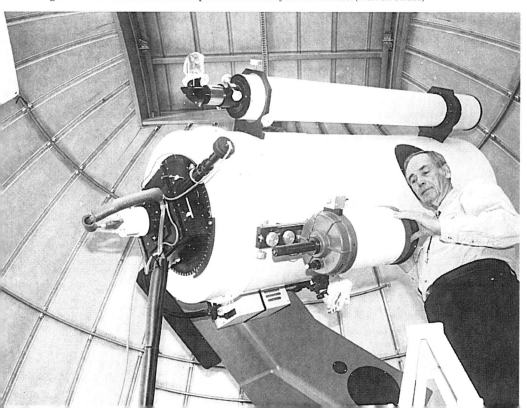

supplied about 50 measurements of Hale–Bopp's position as part of this effort. Looking back, he says he found this a very satisfying experience, as he enjoys projects which can be described quantitatively, i.e. in terms of precise numbers. The Hale–Bopp experience proved interesting enough to make him continue with astrometric work, and having set off along this course the challenge of looking for fainter and fainter objects was irresistible. In a manner that seems almost inevitable to him, but absolutely astonishing to many other people, he found himself making astrometric observations of objects at the very edge of the solar system.

For Warren Offutt a typical observing session starts off with a list of specific objectives for the night. He usually aims to detect and measure the positions of a selection of asteroids fainter than about magnitude 18. Offutt restricts himself to such faint objects since if he does not, he often comes across other asteroids by chance. When this happens his quest for completeness makes him feel obliged to follow them up as well. Before long the number of objects he is trying to track proliferates so quickly that he says it becomes a chore, not a pleasure, to observe them. If the observing goes according to plan, he is usually finished by about 1 a.m. If not he sticks it out all night to get his target list completed. Hopefully, things will go well as Offutt admits that he is not physically up to observing the whole night for several nights in a row. However, like many astronomers a third his age, he still feels bad about stopping observing when the weather is good and says that when he does, his conscience bothers him a little. To him, the tracking down of objects that would have been almost undetectable to even a professional astronomer using the best equipment in the world only a few decades ago is an interesting project with which to test the quality of his telescope and his personal observing skills.

When looking for these very faint objects Offutt usually makes a number of 20–30 minute exposures. During each exposure his telescope tracks the stars using an automatic guider which, after some years of tuning, will keep it on target for hours at a time. Over the course of a night he makes eight or so images and then adds two or three of them together to bring out the faintest objects. Adding in more than three images does not help. The motion of even a very distant asteroid is enough to move it noticeably after an hour or so and adding more frames does not improve the object's detectability unless the frames are shifted to allow for the object's motion before being added. Shifting and adding images in the reference frame of the aster-oid is a common trick for professional astronomers who have large computers, professionally written image analysis packages and full-time software engineers to help them, but Offutt only does it in very special cases. As he puts it, 'The extra processing increases the labour greatly and I have to do everything here, including vacuuming the floor'. When not vacuuming, Offutt was instrumental in providing recovery observations for a number of distant objects and as a tribute to his dedication to this most arcane of observing tasks, Minor Planet number 7639 was named Offutt in his honour.

The success of groups discovering objects using what might be described as medium-sized telescopes encouraged others to try their luck with bigger instruments. The objective of these searches was to find smaller, and so fainter, trans-Neptunian objects to establish how the number of small objects compared with the number of large ones. This information, which is called a size distribution, can help reveal the history of the Kuiper Belt. It may tell astronomers if the objects in the trans-Neptunian region are still growing or if, like the main belt asteroids, they are now slowly grinding themselves down into dust. A few groups decided that instead of searching large areas of sky for a few bright objects, they would concentrate on going very deep over small areas to see how many really faint objects turned up. Depending on the details of the size distribution, this can actually be quite a good way of finding things since, for most populations, there are a lot of small objects for every large one. So, provided the search goes deep enough, the smaller area searched will be balanced by a larger number of potential discoveries. Since to go deep requires staring for long periods at a single location rather than covering a wide area, these surveys are often called 'pencil-beam surveys' as they are akin to sticking a pencil through the sky to see what can be found.

Several pencil-beam searches have been conducted by Brett Gladman and colleagues using both the 3.6 m Canada–France–Hawaii Telescope (CFHT) on Mauna Kea and the 200 inch (4.8 m) Hale Telescope at Mt Palomar. Gladman, who at the time was based at the Canadian Institute for Theoretical Astrophysics (CITA) in Toronto, was interested in the problem because he realised that it was possible to push the detection of faint objects to the limit by using large tele-scopes, sensitive detectors and novel observational techniques. Gladman felt that the searches being carried out by Jewitt, Luu and others in the mid 1990s were too confined in the range of magnitudes which they were finding to reveal much about the size distribution of the trans-Neptunian population. So, since wide-field cameras able to scan huge areas in search of the rare, large objects were not yet available, he set about finding the smallest objects possible.

One search was done from Mt Palomar in September and October 1997. A total of five nights of observing time with the 200 inch Hale telescope were committed to just three fields, each of a square of a little less than 10 arcminutes a side. Since the survey was aimed at the faintest detectable objects, the searching was not done by blinking pairs of images. Gladman used a two-stage process which began by adding together all the frames of each field using a variety of motion vectors expected for typical trans-Neptunian objects. These combined frames were then searched for any evidence of faint point-like sources amidst the forest of trailed star and galaxy images smeared out by the shifting process. Simple visual inspection of the images was not enough. Gladman and his friends found that the best way to search was to blink a series of different images of the same region which had been combined using different apparent motion vectors. When they did this they found that real objects showed a very distinctive pattern, they got brighter as the frame with the most nearly correct apparent motion vector was approached and then faded away afterwards. Once something had been found, or its presence suspected, the process was repeated using a finer range of motion vectors. Eventually, the correct rate of motion was identified and the moving object became a well-defined point source in the image.

The strategy they chose was to observe close to opposition and to pick fields which were relatively empty. Choosing empty fields minimised the risk that the objects they were seeking would be lost in the glare from bright stars. They also chose fields which they knew contained a previously identified trans-Neptunian object. Although the presence of the known object could not be counted for statistical purposes, it could be used to check their data reduction technique. Measurements of the known object's position could also be used to update the astrometric database of observations for orbital calculations. The first field they chose was one near an object designated 1996 RR_{20}. It turned out to be a happy choice as blinking the images at the telescope immediately revealed a new bright candidate, subsequently designated 1997 RT_5. Coincidentally Alan Fitzsimmons independently discovered this object the same night. Fitzsimmons was at the Isaac Newton Telescope in La Palma conducting a programme of

Figure 5.3 Four faint candidates are visible in this image taken as part of Brett Gladman's search for small trans-Neptunian objects. (Brett Gladman.)

astrometric follow-up of known objects and was observing 1996 RR_{20} for this very purpose. Gladman located two other faint objects when the shifting and adding technique was applied to his data. One of these two was subsequently recovered in October. The recovery observations were made by Elanor Helin and Dave Rabinowitz, the discover of Pholus, who had by now left Spacewatch and moved to California. The other object, detected with a magnitude of about 25.6, was not recovered and has now been lost.

A second field was targeted on the position of a relatively bright object designated 1996 TO_{66}. As well as this fairly bright object, the images also revealed a much fainter unknown one with an R (red) magnitude of about 25.8. It was designated 1997 RL_{13} and was at the time the faintest object ever given a minor planet designation. The object is probably only 40 kilometres in diameter and was about 44.5 AU from the Sun at the time of its discovery. It has not been seen again. The third field, which was observed under rather poor conditions in October, did not reveal any new objects. Observations from the Canada–France–Hawaii Telescope in April did not go as deep, but they did reveal one new object on just a single night. Normally, a single night observation would be of little use, but the same field was observed by Alan Fitzsimmons' group a few days later. Fitzsimmons' observations allowed the object to be identified on several other images and permitted the designation 1997 GA_{45} to be assigned. Since then Gladman has moved to the Observatory of Nice on the south coast of France and has found a number of still fainter objects. An observing run on the CFHT in February 1999 led to seventeen new objects being catalogued. One of these was designated 1999 DG_8 which, with an R magnitude of 26.5 and a distance of 62 AU, became both the faintest and the most distant trans-Neptunian object to receive a minor planet designation. Another observing run in January 2000, also at the CFHT, revealed two more very distant objects (2000 AC_{255} AC and 2000 AF_{255}) both about 53 AU from the Sun. In addition to continuing his searches, Gladman has also been trying hard to recover some of the very distant objects discovered by himself and others.

Similar deep surveys were carried out by Jane Luu and Dave Jewitt over a range of nights between 1994 and 1996 and by Caltech astronomers E. I. Chiang and Micheal (Mike) Brown in 1997. These observations were made with the 10 m Keck telescope in Hawaii. Each group detected a few faint objects, but their orbits remain very uncertain. This is hardly surprising since, almost by definition, these deep surveys find mostly very faint objects. Astrometric follow up of these faint objects is almost impossible without large amounts of time on big telescopes, and this is seldom available. It was only a series of coincidences which led to some of the objects in Gladman's survey being observed by other telescopes and such coincidences tend to be few and far between. Also returning to the hunt for faint trans-Neptunians was Gary Bernstein, who with Piet Hut had missed out on finding the first Kuiper Belt object as early as 1991. Bernstein was using data from the same telescope as last time, the 4 m Victor M. Blanco telescope at Cerro Tololo in Chile, but this time instead of the measly 1024 by 1024 pixel CCD available in 1991 he was using a new instrument called the BTC, or Big Throughput Camera. The BTC combines four 2048 by 2048 pixel chips to make a CCD array with a staggering 16 million pixels, able to cover half a square degree of sky. These images were searched using the technique they had used unsuccessfully in 1991, combining several 10 minute exposures of the same field using a range of different motion vectors. This time, however, the larger collecting area of the BTC did the trick; eight new objects were found during three nights of observing.

As more objects were discovered and their orbits confirmed, it became obvious that there was more to the trans-Neptunian region than met the eye. As early as 1994, in IAU Circular 5983, Brian Marsden had pointed out that while the first two objects discovered were in more-or-less circular orbits beyond Neptune, several of the autumn 1993 discoveries were not. The other objects seemed to be in elliptical orbits which came close to, or crossed, the orbit of Neptune.

Figure 5.4 1999 DG$_8$, the most distant trans-Neptunian object yet observed. Several hours of exposures from the 3.6 m Canada–France–Hawaii telescope have been reconstructed to mimic a brief exposure from a telescope 50 m in diameter. This makes the object stand out without being buried by the trailed images of stars in the field. (Brett Gladman.)

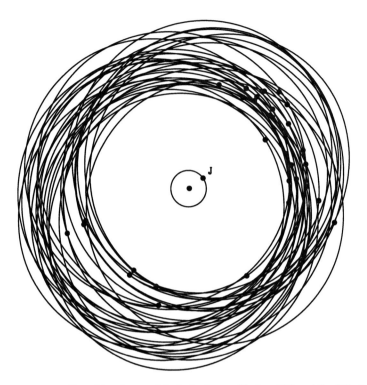

Figure 5.5 The orbits of some of the better known 'Plutinos'. The orbits of Neptune and Pluto are included, but Pluto is indistinguishable from the others. (Chad Trujillo.)

Knowing that close approaches to a planet can change the orbit of a comet or asteroid and send it inwards towards the Sun, or eject it from the solar system entirely, Marsden suggested that the orbits of the autumn 1993 objects were rather special. Specifically, he suggested they were in orbits which took them around the Sun with a period that was an exact fraction of Neptune's orbit. Under certain circumstances this would protect them from being unceremoniously moved into another, far less stable orbit. The resemblance of the situation of these objects to that of Pluto, which also crosses the orbit of Neptune, led Dave Jewitt to refer to then as 'Plutinos' or little Plutos. The realisation that Pluto's orbit was not unique also had another side effect; it eventually started a debate about the planetary status of Pluto itself.

Having defined one class of Kuiper Belt objects as Plutinos, it seemed inevitable that names would be invented for other orbital groupings that were being identified. Apart from the Plutinos, most of the remaining objects were in roughly circular orbits about 42–45 AU from the Sun. This is distant enough that they are almost unaffected

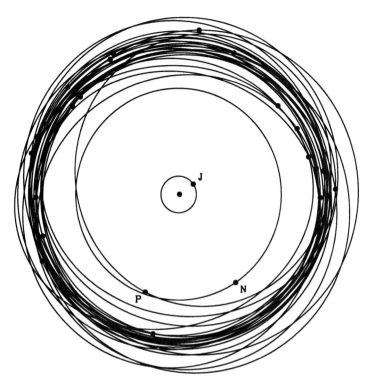

Figure 5.6 The orbits of some members of the Classical Kuiper Belt. These near circular orbits do not cross that of Neptune. (Chad Trujillo.)

by the gravitational forces from the rest of the planets and their orbits are fairly stable. The first object found in such an orbit was 1992 QB_1 so, following an astronomical tradition of naming objects after the first member of each class discovered (e.g. T Tauri stars and BL Lac objects), Brian Marsden suggested that these objects might be referred to as Cubewanos (Q-B-1-ohs). This name did not find much favour amongst the rest of the astronomical community. Instead, and apparently spontaneously, the distant objects began to be called 'Classical Kuiper Belt objects', recognising that they were in orbits which most resembled the predictions made by Kuiper and Edgeworth in the 1940s and 1950s.

An object in a rather different type of orbit was discovered in October 1996 by Luu, Jewitt and his graduate students Chad Trujillo and Jun Chen. Designated 1996 TL_{66}, it was discovered on 9th October at the University of Hawaii's 2.24 m telescope on Mauna Kea using a new, mammoth CCD camera containing no less than eight, 2048 by 4096 pixel, detectors. Built by Gerry Luppino and known as the '8K array' the new camera had an effective size of 8192 by 8192 pixels and a field of view about 18 arcminutes square. Since most of the three dozen or so Kuiper Belt objects which had been discovered up until then were faint, the new detector was being used to a search a relatively large area of sky in the hope of detecting a few brighter objects that might be good targets for detailed physical studies. The huge size of the new array made blinking the entire images impossible, and so they were searched using a computer program developed especially for the task as part of Trujillo's PhD project.

Trujillo's Moving Object Detection Software searched sets of images of a single region of sky which had been taken in fairly quick succession. After the usual processing steps to flat-field the images and determine the sky background, all the objects in each image were identified automatically. Cosmic ray hits and other defects were removed and then the images were aligned electronically. All the stationary objects were then discounted. Finally, the software looked for objects which were starlike, about the same brightness in each image and seemed to be moving with a constant velocity in the range expected for a distant solar system object. The coordinates of any candidate objects which passed these criteria were listed and circles were then drawn automatically on the images, highlighting the regions containing potential candidates. Once the few areas of interest were so marked, the images were then blinked to check the reality of each candidate. This final step was required since the human eye is actually much better at discerning a real moving object from chance alignments of noise than even very advanced computer software. Trujillo estimates that using his software to highlight only the most promising detections for human inspection cuts down the amount of time required to search the plates by what he calls 'an order of magnitude', making a task that would otherwise have taken several hours per frame possible in just a few minutes.

At 21st magnitude, 1996 TL_{66} was the brightest trans-Neptunian object found to date. While initial attempts to determine its orbit suggested that it might be a Plutino, follow-up observations in December 1996 by Carl Hergenrother of University of Arizona were not consistent with such an orbit. Brian Marsden deduced that either there was something wrong with one of the sets of observations or the orbit was rather unusual. Marsden asked Warren Offutt if the Cloudcroft Observatory could obtain further observations to clarify the situation. Offutt once again stepped into the breach left by the telescope

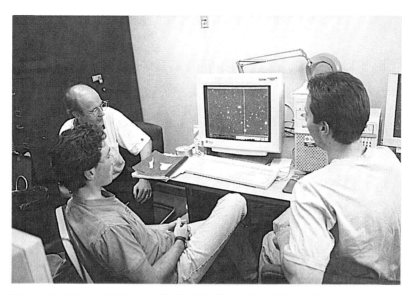

Figure 5.7 Left to right. Dave Jewitt, Chad Trujillo and Scott Sheppard examine an image taken from the 12K CCD camera on the Canada–France–Hawaii telescope. Trujillo's moving object software marks each candidate Kuiper Belt object by drawing a circle around it. (John Davies.)

time-allocation committees. After taking a series of observations on a particularly good night in January 1997, and devoting some 'tender loving care' to drag the faint signal of the asteroid out of the noise in his image, he was able to measure its position well enough for Marsden to solve conclusively the mystery of 1996 TL_{66}. The entire series of observations showed that 1996 TL_{66} was close to the perihelion of a highly eccentric orbit with a period of 788 years. This orbit takes 1996 TL_{66} over 130 AU from the Sun, well beyond the confines of the classical Kuiper Belt. The orbits of 1996 TL_{66}, and a few similar objects discovered a few years later, are highly elliptical. These objects are in or near the classical Kuiper Belt when closest to the Sun, but spend the rest of their time very much further away. Their huge elliptical orbits take them hundreds of astronomical units from the Sun. They are known as scattered disc objects.

Chad Trujillo is not the only person using software, rather than manual blinking, to search for distant asteroids. The Spacewatch project, which in 1992 discovered the second known Centaur, 5145 Pholus, has long been interested in using computer power, rather than brain power, to search its large and ever-expanding datasets. Spacewatch was the brainchild of asteroid astronomer Tom Gehrels, who had long promoted the idea of a telescope dedicated to studying

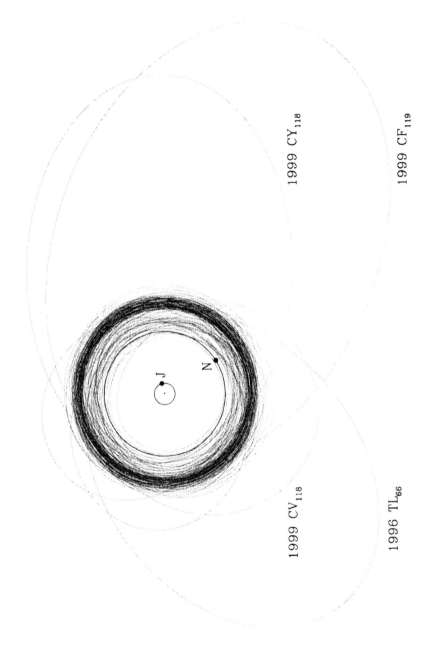

Figure 5.8 The orbits of four scattered disc objects. The Plutino region and the Classical Kuiper Belt are shown together with the orbits of the planets Jupiter and Neptune. (Dave Jewitt.)

the small objects in the solar system. Although the idea for such a telescope had been around since about 1969, the small number of astronomers interested in asteroids meant that it was not until about 1980 that the project began to take shape.

In 1981 a historic 0.9 m telescope became available to Gehrels and his small group. It was the original Steward Observatory telescope, first erected on the university campus in Tucson, Arizona in 1921 and moved to the clearer and darker skies of Kitt Peak in 1962. Spacewatch acquired the telescope and by 1983 was operating it with a 320 by 512 pixel CCD camera. In line with other observatories, Spacewatch rapidly acquired larger CCDs, moving to a 2048 square device in 1989. Since it is a dedicated asteroid search telescope, and has to cover large areas of sky repeatedly rather than going very deep on small areas, Spacewatch operates differently to most of the other groups we have so far encountered. Instead of pointing at a selected region of the sky and then turning to counteract the Earth's rotation, the Spacewatch telescope does not rotate with the sky. Instead, it points at a fixed position and allows the image of the sky to move across the detector. Under normal circumstances this would result in the star images smearing out, or trailing, and would be quite unacceptable. However, the Spacewatch CCD is aligned on the sky so that individual star images move parallel to the edge of the chip. In effect the stars drift along a single row during the exposure. By adjusting the rate at which the charge in each pixel is transferred to the next column during the readout process, the reading of the array is synchronized with the rate of drift of each star across the chip. When the accumulated charge reaches the column at the edge of the chip it is transferred down the end column and out of the array in the usual manner. This readout technique is called drift scanning. It is very efficient since the CCD is constantly exposed to light (there is no shutter opening and closing) and there is no dead time between exposures while a shutter is closed and an array is read out. In the context of the farmer trying to measure the rainfall across the field described in Chapter 4, the buckets are being emptied into each other across the field at exactly the same rate that small clouds are drifting overhead. So when the buckets reach the end of the field, each contains water from just a single small part of the cloud directly above it. Drift scanning for asteroids at Spacewatch began in earnest in 1984.

Spacewatch's automatic Moving Object Detection Program (MODP) came into use in 1985. Its development required about eight

Figure 5.9 The Spacewatch telescope on Kitt Peak, in Arizona. The astronomer is Jim Scotti. (Lori Stiles, University of Arizona.)

man-years of work, mostly by Dave Rabinowitz and James 'Jim' Scotti. Scotti was born in 1960 in a small coastal town called Bandon, in Oregon, where he lived for 8 years. As a boy he was captivated by the sight of astronauts walking on the Moon and became interested in space travel. As the Apollo programme came to an end, his interest expanded to encompass not just the Moon, but also the rest of the solar system. He bought his first telescope with money he had earned by babysitting and his interest in astronomy blossomed. He went to the University of Arizona and was awarded a degree in astronomy in 1982. The very next week he started working for Spacewatch full time, having previously worked there as an undergraduate. He now spends about six nights per month observing at Spacewatch.

During normal observing the Spacewatch telescope is used to scan a strip of sky about half a degree wide and seven degrees long three times each night. The MODP software records the positions and brightnesses of all the objects detected during each scan. When observing is finished, the software compares the three lists and looks for something which appears to be moving in a consistent way from scan to scan. Usually many hundreds of such candidates are recorded. These candidates must then be examined by the observer to decide which are real and should be followed further and which are just chance alignments of noise or bad pixels and can be ignored. This checking takes several hours of the daytime, allowing the observer only a limited time to eat, sleep and prepare for the following night's observing. Of course, the number of false detections can be reduced by making the software's selection criteria more rigorous, but experience has shown that if this done some real detections are also missed. Although only applicable to very rapidly moving objects, and so not relevant to our story, Spacewatch also has software which searches for streaks caused by objects moving so quickly that they trail out during a single exposure. Using this fast-moving-object software Spacewatch has discovered a number of asteroids which pass very close to the Earth.

Although originally associated more with discovering objects passing close to the Earth, the Spacewatch project always had in mind the search for more-distant objects. Tom Gehrels describes the overall goals as discovering small objects throughout the solar system in order to study their statistical and dynamical properties. The discovery of Pholus in 1992 drew attention to Spacewatch's capability to detect more-distant objects and this was emphasised when Jim Scotti discovered three more Centaurs. Although Dave Rabinowitz has moved on, Scotti has remained at Spacewatch. He still has his boyhood enthusiasm for astronomy, but these days he sometimes regrets that during a typical observing session there is never enough time to stay outside and really enjoy his view of the sky.

Since it first went into operation the Spacewatch telescope has scanned a very large area of sky. The huge amount of data it has taken has recently been searched to see if any very-distant objects might lurk there. New software was written to find slow-moving objects undetected by the original MODP software. On 17th March 1999 this new software had its first success, finding an object which had remained hidden in the data since September 1995. This object, 1995 SM_{55}, was re-located from the ground in 1999 and its orbit is now quite well defined. Soon after, six more objects, all with visual magnitudes between 20.6 and 21.5, were located in the data archive. Although some of these have not been observed again, others (including objects designated 1998 SN_{165}, 1998 SM_{165} and 1998 VG_{44}) have all since been re-located and now have secure orbits.

The Spacewatch team continues its search and by mid 2000 it had discovered two new and interesting objects which do not fit into any of the three previously defined groups. These two objects (1996 GQ_{21} and 1999 TD_{10}) have orbits which sweep across those of some outer planets like the Centaurs, but which are not confined to the main planetary region. At their furthest from the Sun these objects travel deep into space, reaching distances of over 140 astronomical units away. These large perihelion distances are more like those of the scattered disc objects. Not long after these discoveries the Minor Planet Center merged its lists of Centaurs and scattered disc objects. This decision reflects the fact that there are probably no fundamental differences between the two classes of objects, other than that they have been gravitationally scattered in different directions.

Sorting out the dynamics

As early as spring 1994 Brian Marsden had remarked that some of the newly discovered trans-Neptunian objects had orbits that were similar to that of Pluto, but what is it that makes Pluto's orbit special? The general details of the ninth planet's 246-year orbit had been established within a year of Tombaugh's discovery and two things made it stand out from the other planets. Pluto's orbit is the most highly inclined, and the least circular, of all the planets. In fact, the orbit is so eccentric that it crosses the orbit of Neptune. How can Pluto survive in such an orbit?

Ancient philosophers believed that the Earth was the centre of the Universe and that the planets moved in circles. This geocentric model survived until Nicolas Copernicus proposed a heliocentric (Sun centred) model of the solar system in the early sixteenth century. Although he was correct in his belief that the Earth went around the Sun, Copernicus erroneously assumed that planetary motions must be based on circular forms. The last vestiges of heavenly perfection were removed the following century when Johannes Kepler showed that the orbits of the planets are not circular, but elliptical. However, the amount by which the orbits of the major planets differ from a circle is generally quite small, so much so that if a diagram of the solar system is drawn to scale on a normal-sized piece of paper, the average eye would be hard put to detect the elliptical nature of most planetary orbits. Mathematically, the shape of an ellipse is denoted by its eccentricity 'e', and to avoid getting embroiled in the mathematics of conic sections (circles, parabolas and hyperbolas) all that is important to know is that the larger the value of e, the more squashed is the ellipse. The eccentricity of a circle is zero; for something the shape of an egg it is about 0.5. Most of the planets, from Venus to Neptune, have eccentricities of less than 0.06,[†] and so their orbits are indeed almost circular. However, with an eccentricity of 0.254, Pluto's orbit is quite elongated and its heliocentric distance ranges between 29.6 and 48.9 AU. Neptune has an almost circular orbit with a mean heliocentric distance of a little over 30 AU. This means that for about 20 years per orbit, Pluto is actually closer to the Sun than Neptune. This situation is unique; Pluto is the only planet to cross inside the orbit of another.

The existence of Pluto in a Neptune-crossing orbit is surprising. At first sight it would seem highly likely that at some time in the last few billion years Pluto must have been crossing the orbit of Neptune at just about the time Neptune was close to the crossover point. Neptune is much more massive than Pluto, so the effect of such an encounter on the larger planet would have been negligible, but the resulting gravitational perturbation on Pluto would have been drastic. The inevitable encounter should have moved Pluto into a new orbit, perhaps sending it towards the Sun or ejecting it into deep space. The reason why this has not occurred, and why Pluto has survived so long, is that Pluto's 245.96-year orbital period is almost exactly one and a half times that of Neptune's 164.7-year journey around the Sun. This situation, in which two objects have orbital periods that are simple multiples of each other, is called a commensurability. In this case, the relationship means that every time Neptune completes three orbits, Pluto makes two. Consequently, at the end of this little celestial dance the two planets come back into the same relative positions. Five hundred or so years later, Pluto has made two more orbits and Neptune another three, so once again the planets are back where they started. If this special dynamical situation survives despite evolution of the orbits with time, it is called a mean motion resonance. In this case we say that Pluto is in Neptune's 2:3 mean motion resonance. So, provided Pluto is in an orbit that happens to avoid close encounters with Neptune, it will continue to avoid them on a regular basis. As it happens, when Pluto is close to perihelion, and so is crossing Neptune's orbit, Neptune is 90 degrees away around its orbit. This huge distance is easily large enough to prevent a significant perturbation of Pluto's orbit by Neptune's gravity.

Now such an exact commensurability is a very special situation. It seems unlikely that one should occur, and even less likely that it could survive over the age of the solar system. Over millions of years the

---

[†] The innermost planet Mercury has an orbit with eccentricity of 0.2.

weak gravitational influences of the other planets would be expected to add up and disturb this special relationship. Indeed, the Pluto–Neptune commensurability could not exist in the real solar system for very long but for the fact that there are other factors at work. If Pluto is not in an exact commensurability, then over time its orbit will change so that the crossover point drifts closer to Neptune. When this happens Pluto begins to feel an increasing gravitational tug from the much larger planet. The effect of this tug is to change Pluto's orbit by a small amount and to alter the orbital relationship between it and Neptune. The effects of these little perturbations will eventually send Pluto's orbit back the way it came. When this happens Pluto's orbit will begin to move back towards a condition of exact 2:3 commensurability. However, Pluto never manages to stay in this condition since the orbit overshoots the stable point and begins to drift towards that of Neptune from the opposite direction. This means that Pluto's orbit is now drifting towards Neptune from the other side and the gravitational influence of Neptune is now in the opposite sense. Eventually, these perturbations will reverse the drift again. This will send Pluto's orbit back towards the exact resonance which it will reach and then overshoot. Pluto will then begin to repeat its complex orbital wanderings. Pluto never manages to remain quite in the 2:3 commensurability; instead, its orbit wanders back and forth, or librates, around it with a period of about 70 000 years (a little less than 300 Pluto revolutions). So, while the fine details of Pluto's orbit change slightly over millennia, with small variations in inclination, eccentricity and semimajor axis, these parameters oscillate gently around some typical values. Fortunately for Pluto, they never undergo any dramatic changes severe enough to jolt the planet into an unstable orbit from which it can never return to the mean motion resonance.

The first two trans-Neptunian objects found were on fairly circular orbits well beyond the immediate gravitational influence of Neptune, but this was not true for the next four objects discovered. The preliminary orbits for 1993 RO, RP, SB and SC all seemed to indicate that they were substantially closer to the Sun than 1992 QB_1 and 1993 FW. All four seemed to be at heliocentric distances of around 30–35 AU, putting them right in the danger zone for potentially disastrous gravitational encounters with Neptune. Since it was unlikely that many objects could survive in this region for long, it was not long before the idea that they might be in stabilising resonances surfaced. An early hint of this came with the discovery announcement of 1993 RO, with

Brian Marsden noting that the object was located about 60 degrees away from Neptune.

The significance of this rather cryptic remark is that there are a number of potentially stable points connected with giant planets. The best known examples of these are the regions occupied by the so-called Trojan asteroids, the first of which was discovered in 1906. It was named 588 Achilles, a character from the *Iliad*, Homer's epic poem about the Trojan Wars. Achilles was found to have a mean distance from the Sun of about 5 AU, placing it at a very similar heliocentric distance as Jupiter. This seems to be a very unlikely place to find a four and a half billion year old asteroid, but Achilles was able to survive because the gravitational influences of the Sun and Jupiter combine to make the specific region of space it occupies fairly stable. The existence of five gravitationally stable points in a system containing two massive bodies and a third one of negligible mass was pointed out by the eighteenth century mathematician Joseph Louis Lagrange, and the stable points are named Lagrangian points in his honour. Three of the Lagrangian points lie on a line connecting the two bodies. The two others fall 60 degrees ahead of and behind the smaller of the two large masses. As other asteroids were discovered in orbits similar to that of Achilles it became traditional to name them after characters from the *Iliad*. The leading group contains characters from (mostly) the attacking Greek army and the trailing group is populated by the defenders of Troy. Because of the names given to the objects that inhabit them, these stable regions are often referred to as Trojan points. However, since the real solar system contains more than two planets and one asteroid, the Lagrangian points are not points at all, but regions. They are 1:1 mean motion resonances of Jupiter.

Trojan asteroids are not fixed in position as rigidly as some astronomy books suggest. They librate around their Trojan point and can wander a fair way from it before drifting back. However, on average, they are found in a region close to the magical 60 degrees from the planet. The implication of Marsden's remark was clear, was 1993 RO a Trojan asteroid of Neptune? The problem was that with such a small observed arc, the orbit of 1993 RO, like those of the other new discoveries, was quite uncertain. What was needed was more observations, preferably a few months hence, when the new objects had moved somewhat and the observations could provide a bigger arc. However, there was a Catch 22. To improve the details of their orbits, the objects had to be observed when they had moved a fair way across the sky, but without an orbit to predict their positions, could they ever be found again? Luckily there was a clue. Dave Tholen had observed 1993 SC in November 1993, and while his positions still allowed the object to be in a variety of orbits ranging from 34 to 44 AU, Brian Marsden suspected that, like Pluto, 1993 SC was in resonance with Neptune.

During the winter of 1993–4, the movement of the Earth meant that the new objects slipped behind the Sun and for several months they only rose above the horizon during daylight. This made recovery observations of them impossible. It was not until early summer of 1994 that attempts to relocate them could begin. In May, Brian Marsden published a set of predicted positions for the four objects based on his assumption that their orbital stability was assured by them being in 2:3 resonance with Neptune. These positions relied heavily on this assumption and could have been in considerable error, especially if the objects were in fact Trojan asteroids of Neptune.

The searching season opened in about June as the objects became visible in the morning sky just ahead of the dawn. For a while there was no news of them. This ominous silence was broken in the late summer with the recovery of 1993 SC and 1993 RO from two observatories. In August, Mark Kidger found the two objects using the Isaac Newton Telescope in La Palma and Dave Jewitt and Jane Luu picked them up in September from the 10 m Keck telescope on Mauna Kea. Later in September, the Hawaii team recovered 1993 SB from the UH 2.24 m telescope. Although not sufficient to tie down the orbits of these objects with very great precision, the new observations were all consistent with orbits in the 2:3 Neptune resonance.[†] This fuelled speculation that there might be a much larger population of similar objects in Pluto-like orbits. According to Marsden, it was the realisation that there were other objects in resonance with Neptune that finally allowed Pluto to 'make sense'; this was the key to understanding the whole trans-Neptunian region.

The fourth object, 1993 RP, was never seen again, but the existence of a population of objects in 2:3 resonance with Neptune was soon confirmed. Note that the orbits of these objects are not all the same, nor are they exactly the same as Pluto. The orbits have a range of eccentricities and semimajor axes and the majority of them do not

---

[†] None of the four turned out to be Neptune Trojans even though such locations should be stable.

actually cross the orbit of Neptune (1993 SB is one of a few known objects which do). The point is that they all come close enough to the orbit of Neptune to have unstable orbits but for the effects of resonances, which prevent a close encounter with the planet itself. The existence of this population of Plutinos begged the question of whether or not other of Neptune's mean motion resonances were also populated. Theory indicated that there were other resonances; in particular there should be a population of objects with periods of close to three hundred years in the 1:2 resonance. This resonance lies about 47.5 AU from the Sun, so objects in it will be fainter and harder to find, and it was not until 1996 that the first such object (1966 TR_{66}) was actually discovered.

However, the discovery of numerous objects in mean motion resonances did not mean that all the trans-Neptunian objects were so stabilised. In particular, the orbits of the first two objects found, 1992 QB_1 and 1993 FW, were rather more distant than the Plutinos. Even without the protection of resonances, they never approach Neptune closely enough to be strongly perturbed by its gravity. These two objects, and others like them, most closely resembled the population of objects described by Edgeworth and Kuiper and so this region of space became known as the classical Kuiper Belt. However, there remained the 'dirty little secret' of the Kuiper Belt; if this was the source region of the short-period comets, what actually caused objects to escape the belt and move into the inner solar system?

This problem was not one that could be solved directly by observation. In astronomy things usually happen very slowly and there is no chance that any of the presently known trans-Neptunian objects will evolve into comets during the lifetime of anyone who is observing them today. Astrophysicists have similar difficulties when they deal with the evolution of stars and galaxies; here too, things usually happen too slowly to follow during a single lifetime. Luckily for the astrophysicists there are other ways of tackling their problems. Star formation is going on throughout the galaxy and different star-forming regions are of different ages. By observing different regions astronomers can study snap-shots, if not moving pictures, of the star-forming process. Astronomers studying extragalactic objects observe objects so far away that the finite speed of light means that when they observe distant galaxies they are peering backwards in time as well as across vast reaches of space. This means that cosmologists can study galaxies with a range of ages to try and understand how they form and evolve. Solar system astronomers do not have either of these luxuries. They are studying just one solar system (the recently discovered extrasolar planets seem to be in systems rather different to ours) and the distances are tiny by comparison with usual astronomical standards. The look-back time to the edge of the solar system is measured in hours, not billions of years. If astronomers want to peer into the distant past of our solar system, or project events far into its future, their only recourse is to model it mathematically.

Since the effects of gravity are well understood, there is nothing conceptually difficult about tracing the motion of one object under the influence of another. In principle it should also be easy to add in further objects and see how mutual gravitational interactions affect the overall picture. The problem is the sheer number of calculations which need to be performed. As soon as one body moves, its effect on all the others changes. Analytical techniques have been developed over the centuries to get around this difficulty, but these are usually restricted to just a few objects. For example, analytical methods can investigate the evolution of a couple of planets moving around a star, but they cannot deal with a huge ensemble of particles all moving about at the same time. It is only with the advent of computers able to make many calculations in microseconds that detailed mathematical modelling of the evolution of the solar system has been possible. Modern computers allow theoretical astronomers to simulate the motion of objects as they move through space under the influence of the combined gravitational forces of the planets and to determine the effects of the competing gravitational tugs on the evolution of these orbits. This process is known as integrating the orbital elements to see what they will look like some time in the future, or what they were some time in the past. Integrating orbital elements for real objects needs to be done so their positions can be calculated accurately enough for observations with large telescopes (it is especially important if the object happens to have passed near another planet recently and may have had its orbit changed significantly during the encounter) but orbital integrations can be used as a research tool in their own right.

One approach is to take the details of an object that actually exists and to make some small, but significant, change in one aspect of its orbit before allowing the computer to track how it might then evolve. For example, take the orbit just as it is, but move the object along its path a little. This will make it seem as if one object had suddenly jumped forward a few years while everything else stood still. The effect of this can vary from not very significant, which suggests the orbit is fairly stable, to quite dramatic. For example, Iwan Williams of Queen Mary and Westfield College in London, one of the group which had discovered 1993 SB and 1993 SC, considered the evolution of the orbit of 1993 SB over a ten million year period. He found that even though it crossed the orbit of Neptune, and its orbit did oscillate gradually with time, the orbit was stable during this period and never changed significantly. However, if he kept all the orbital details the same and just assigned them to a date five years hence (and remember this is only 2% of 1993 SB's 250 year orbital period) the result was rather different. The main effect was to change the geometry when 1993 SB made its closest approaches to Neptune. The effect of this was to cause very significant changes in all its orbital parameters. While the 1993 SB 'clone' survived the ten million year integration, it was clear that something dramatic was going to happen to it, and probably sooner, rather than later.

This result reveals an important point about the stability of these resonant orbits; while objects in them might well be quite stable, even slightly different orbits may be unstable. What we see today is not an intact population of ancient objects, but just the lucky few survivors that happened to find themselves in stable orbits, or were somehow pushed into them before they could be ejected from the solar system. This is why Pluto has survived to this day. It is not just a fortunate coincidence that its orbit is in a mean motion resonance with Neptune; if it had been otherwise Pluto would not have survived long enough for Clyde Tombaugh to discover it.

Another way of using computers to study solar system evolution is the one taken by theoretical physicists like Martin Duncan and Hal Levison, who we last encountered in chapter 4 unsuccessfully trying to detect the Kuiper Belt from a telescope in Arizona. They, and others, have developed very detailed computer codes to follow the evolution of objects initially in orbits close to Neptune. They can then compare the outcome of their simulations with what is actually observed out there in real, as distinct from cyber, space. To do this, they set up models in which hundreds, or sometimes thousands, of test particles (essentially theoretical Kuiper Belt objects), with a wide range of orbital parameters, are set in motion. Their orbits are then integrated over very long periods. The details vary from simulation to simulation, depending on the objective of each experiment, the cleverness of the code written to do the calculations and the amount of computing power available. For example, in experiments conducted around 1992, the orbits of 1000 test particles were integrated at 1 or 10 year intervals over a period of 1000 million years. This is equal to about one quarter of the age of the solar system and should have given a reasonable idea of how the final result would come out if there had been time to allow the simulation to run longer. Of course, some approximations had to be made – the test particles were assumed to be without any mass, so mutual interactions could be ignored. Also, not all the planets were included, as the influence of the gravity from the small and distant inner planets is tiny compared with that of the four giant planets. In fact, when studying objects beyond Uranus, even the effects of Jupiter and Saturn can often be ignored, provided their effects on the orbits of Uranus and Neptune are taken into account before calculating the gravitational influences of Uranus and Neptune on the test particles.

No computers yet exist that can integrate objects in every conceivable orbit over the age of the solar system with high enough temporal and spatial resolution to produce any kind of exact answer. Even if there were such machines, it would still not be possible to complete the job. The effect of non-linear dynamics, or chaos theory, makes it impossible to predict things in the real world with complete accuracy. To see why this is so, consider that in a real solar system one of the particles might be hit by a meteorite which could change its orbit very slightly. This minor incident might not make any difference to the outcome, but it might cause a slightly different effect on the path of another particle. This in turn might have knock-on effects elsewhere. Before long, astronomically speaking, these effects could ripple through the entire population and produce an outcome quite different from the one which would have resulted if the first particle had not had that meteorite impact. Such subtle effects cannot be fully modelled, so infinite precision is never going to be possible in these dynamical simulations.

Since it is impossible to do everything, detailed studies of the effect of specific parameters on the outcome of a simulation are done by starting with test particles having a restricted range of some parameters and a complete mixture of all the others. For instance, to investigate the effect of semimajor axis on the outcome of a test, a selection of orbits all having fairly low inclinations and a range of eccentricities might be chosen. To see what happens to particles at high inclinations, the initial set might include only a small range of eccentricities, but a wide range of inclinations. Once these initial values have been chosen, then the particles can be put into these orbits at random positions, the orbits distributed around the Sun and the simulations started. In most simulations, once a test particle has reached some critical limit, for example when it comes close enough to Neptune's orbit to have a dramatic encounter, or it has moved more or less out of the gravitational reach of the planets, it is removed from the simulation to save computing power.

This is a very different kind of astronomy from that practiced by Jewitt, Luu, Offutt and the other observers of the real Kuiper Belt. It does not involve travel to exotic destinations such as Hawaii, Chile or the Canary Islands and it does not even necessarily involve staying up at night. It is a job where the tools are computer workstations not telescopes and where capability is measured in terms of processor power, not mirror size, and Megaflops (millions of floating point operations per second) rather than fields of view and limiting magnitude.

Martin Duncan has been very active in this esoteric field and he is now a professor of physics at Queens University in Kingston, Ontario. Like many university academics, he fits his research in around teaching and general administration of his department. His day often starts early, with a check on whatever computer runs are going on at that moment. His group tends to set up long simulations, which may take weeks or months to finish, but he says that it is nice to be reassured that everything is running smoothly. Some of his simulations involve dozens of machines and many months of computing time to build up the statistics of how the virtual solar systems are forming and evolving. While, as he puts it, his workstations are just sitting there doing lots of simple calculations over and over again and probably getting very bored in the process, Duncan has time to think about what it all might mean. He has to consider what further tests need to be done and to try and have his next good idea ahead of the competition. Like the observers who have to compete for telescope time, the theoretical physicists are also competitive people. While they recognise and respect the good work of others, they also sometimes knock themselves on the head and ask themselves, 'Why didn't I think of that?'.

What the computer simulations of Martin Duncan and other people doing similar projects have shown is that the structure of the region around Neptune is very much more complicated than originally thought. Objects in near-circular orbits can appear to be perfectly stable for hundreds of millions of years, with just gentle, cyclical changes in their orbital elements over ten million year timescales, then suddenly the gradual build-up of the gravitational influences of the planets conspire to produce a rapid change. Over a few million years, the once apparently stable orbits can change rapidly until they become Neptune crossing, or at least get close enough to Neptune to suffer a major gravitational perturbation, and be flung out of the trans-Neptunian region completely.

By studying the fate of thousands of ghostly objects, Hal Levison and Martin Duncan have been able to map out the likely structure of the trans-Neptunian region. It is a strange kind of map. It does not plot positions in space, or motion around the Sun, but islands of stability in the strange mathematical space of orbital elements. An example of one of these diagrams is shown in figure 6.1. This shows that the trans-Neptunian region is expected to have a complex structure, although one whose general features are fairly easy to understand. In general, any object in an orbit that brings it to within about 35 AU of the Sun will be removed by the cumulative effects of Neptune's gravity quite quickly. The models show that this region is expected to be fairly empty. Any objects found in such orbits today must have drifted into them quite recently and are already doomed. For them, a dramatic gravitational encounter with Neptune and permanent removal from the Kuiper Belt is only a matter a time. The only exceptions to this rule are objects such as the Plutinos whose orbits are librating around Neptune's mean motion resonances. They appear to be stable for the age of the solar system.

Moving outward, there is a region from about 40 to 42 AU in which orbits are highly unstable. This is a region where another kind of resonance, a so-called secular resonance, becomes important. We saw that for a mean motion resonance to occur the orbital periods of two objects need to be a simple ratio of each other, like the 2:3 relationship between Neptune and Pluto. If the orbits of the planets remained fixed in space, then that would be all there was to worry about. However, in reality, the mutual gravitational interactions of the planets cause their orbits to precess. Precession is a process in which the direction defined by the long axis of an elliptical orbit drifts slowly around the Sun. It is similar to the way in which the hoop of a hula hoop dancer drifts gradually around the dancer's waist. Provided the rates of precession of any two orbits are not the same, then an

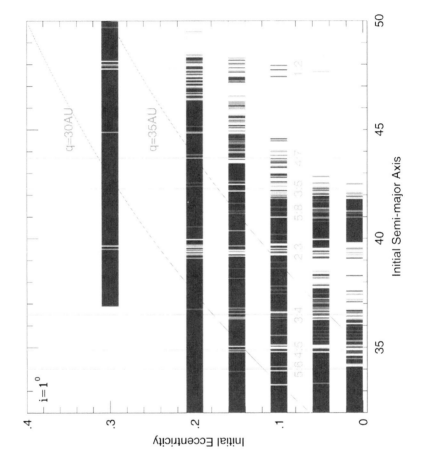

Figure 6.1 This figure shows the predicted dynamical lifetimes of trans-Neptunian objects with various initial orbital characteristics. The original orbits of the test particles are defined by some combination of initial eccentricity and semi-major axis and the simulation is allowed to proceed. The curved lines marked by q=30 and q=35 show the combinations of eccentricity and semi-major axis which allow objects to come within 30 and 35 AU of the Sun at their closest approach (perihelion). Note that almost no orbit within (to the left of) the q=30 AU line is stable over the age of the solar system. Various Neptune mean motion resonances are marked by vertical lines, such as the 1:2 resonance at about 47.8 AU. The stable (light coloured) regions tend to be close to these resonances.

object outside a precessing planet experiences a gravitational force which is symmetrical. Over a suitably long timescale the changing gravitational influences of the interior planet average out. If, however, the period of precession of one object is the same as that of another, then the situation is not symmetrical and there can be significant gravitational torques on the smaller of the two objects. The effect of these forces can be to change the eccentricity of the smaller object's orbit. In a solar system like ours, in which all the bodies are not orbiting in exactly the same plane, additional secular resonances can occur. Each planet's orbit crosses the invariable plane of the solar system at two points, called the ascending and descending nodes. The position of these nodes also precess and if the rate of precession of the nodes of two orbits are the same then gravitational forces can change the inclination of the smaller body's orbit. These types of resonances are normally denoted by the greek letter ν or ν_1 respectively, followed by the number of the planet doing the perturbing. The secular resonances with Neptune, which has the dominant effect on the trans-Neptunian region because of its proximity, are denoted ν_8 and ν_{18}.

It so happens that the region around 40–42 AU is where several of the secular resonances of Neptune and Uranus occur, and these conspire to make this region unstable. Any objects which have the misfortune to wander into this part of the solar system are rapidly removed. The secular resonances also destabilise objects in high inclination orbits with semimajor axes close to that of Pluto, explaining why the stable Plutinos are not far from the plane of the solar system. Beyond about 42 AU these secular resonances cease to be important and a fairly stable region of orbital space exists. Objects here can often survive for the age of the solar system without needing to protect themselves by remaining in mean motion resonances of Neptune.

Using their computer models the theoretical astronomers, or 'orbital mechanics'[†] as they are sometimes jokingly referred to by observational astronomers, were able to paint a self-consistent picture of the space beyond Neptune. Their calculations showed that it is a region which has been dynamically sculpted by aeons of gravitational interactions into a complex structure. Some regions were expected to contain many objects while others should be almost empty. However, this was all theory; the proof of the pudding was to be in the eating.

[†] Strictly, they are orbital mechanicians.

Would the theory survive a head-on crash with real observations? To find out, astronomers plotted the real trans-Neptunian objects onto the map of stable regions to see how well the actual distribution matches the predicted one.

The answer, which is shown in figure 6.2, is that it fits quite well. Around about 39 AU there is a population of Plutinos with low-inclination orbits stabilised by the 2:3 mean motion resonance. The 40–42 AU gap, which theory predicts should be cleared by secular reso-nances, is indeed empty. Beyond 42 AU are a number of objects spread more or less evenly across the predicted stable region. Only a few objects are seen beyond 45 AU, but this may be a selection effect. At these large distances objects will be that much fainter and corre-spondingly harder to detect. There is, however, one anomaly. Theory predicts that there is a stable region for low-inclination orbits with semimajor axes in the range 36–39 AU. This is comparatively close, closer than the Plutino zone, and so it should be quite easy to find objects there, but in fact none have been discovered so far. How can this be explained?

Figure 6.2 The orbital distribution of real trans-Neptunian objects. Note the concentra-tion of objects in the 2:3 mean motion resonance. These are the Plutinos. Pluto itself is marked by an X just above the line representing objects with a perihelion distance (q) of 30 AU. (Dave Jewitt.)

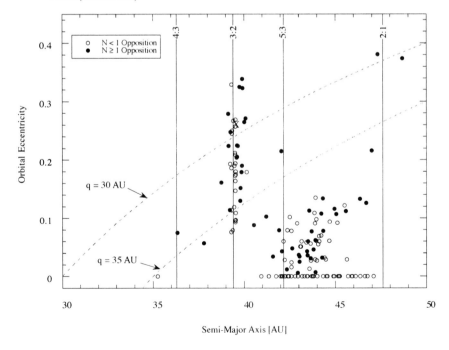

There are at least two possibilities. Orbits in this region are only stable if they are fairly circular, i.e. they have eccentricities of less than about 0.05, and have low inclinations. If some process caused the orbits in this regions to be systematically altered, to have either larger eccentricities or higher inclinations, then the objects would have been ejected by now and the region would indeed be empty. So perhaps the eccentricities and/or inclinations of objects in this region were 'pumped' up to higher values by mutual interactions between the objects themselves. The presently known trans-Neptunian objects are not massive enough to do this, but this inner region may once have contained a few larger objects. This is not by any means impossible; Pluto, its moon Charon and Neptune's largest satellite Triton, are bigger than typical trans-Neptunian objects and all are found in this vicinity. Perhaps a few large objects once existed here and were responsible for gravitationally clearing the region before being removed by collisions or some other as yet unexplained process.

If the idea of creating a few large objects which can clear out the 36–39 AU region and then themselves vanish does not appeal, there is another less *ad hoc* explanation. This hypothesis might bear not just on the clearing of the 36–39 AU region, but on the trapping of Pluto and the other objects in the 3:2 resonance. Perhaps the structure of the solar system we see today is not the same as it was early in its history. In particular, it is possible that changes in the orbit of Neptune might have played a role in sweeping out the 36–39 AU region and trapping objects in mean motion resonances. This sort of effect would not be found by simulations which assume that the orbits of the giant planets are the same now as they were over four billion years ago. If the planet's orbits once moved around the solar system, then the effects of this evolution will be seen in the present structure of the trans-Neptunian region. Studying the Kuiper Belt today can provide important clues to how the outer solar system formed.

Can planets' orbits migrate across the solar system? Yes they can and what is more they probably did. During the final stages of the formation of the solar system a situation must have existed when the four largest planets had more-or-less finished growing and were already on fairly circular, low-inclination orbits. Any nebular gas which had not been incorporated into planetesimals had long since vanished, but large numbers of planetesimals were still to be found in the space between the planets. They were particularly common in the region beyond Neptune where no massive planet had formed. One by one, these planetesimals would have encountered Neptune and, while a small fraction would have crashed into the planet, the majority would have been scattered into new orbits by Neptune's gravity. However, gravitational scattering is not a process that only affects the planetesimal. The principle of conservation of angular momentum means that the orbit of the scattering planet must also be affected. For a single interaction, the effect on the orbit of the giant planet is negligible, but if many encounters are occurring the effects can build up over time. The combined effects of many such interactions may significantly change the orbital energy and angular momentum of the scattering planet.

Early during the scattering process, when there were many planetesimals, there would have been an equal number of objects scattered inwards and outwards. The effect of these events on the scattering planet's orbit would cancel out, leaving no net change in its energy. Later on, this situation would change. The ultimate fate of the scattered planetesimals depended on which direction they went after their first encounter with Neptune. Most of the objects scattered outwards from Neptune could return to encounter the planet again in the future, but this is not true of objects sent inwards. As shown by Ed Everhart, objects scattered inwards are likely to fall under the gravitational control of the other giant planets and eventually reach the sphere of influence of massive Jupiter. If this happens there is a good chance that Jupiter's gravity will eject them from the solar system entirely. So, late in the scattering process, most of the objects interacting with Neptune are the survivors of objects thrown outwards on their first encounter and returning Sunwards. There are comparatively few objects interior to Neptune diffusing outwards. Towards the end of the process the energies of the objects reaching Neptune are biased, the average energy of each gravitational interaction is no longer zero. The encounters are now, in effect, a drag on Neptune and this acts to increase the orbital radius and causes the planet to drift outwards. Uranus and Saturn, which are also outside Jupiter, would have suffered similar, although smaller effects. They would also have migrated outwards slightly. Jupiter, the most massive planet, having indirectly provided the energy to move the other planets outwards by ejecting many planetesimals entirely, would have moved inwards to conserve the solar system's total angular momentum.

These changes in the diameter of Neptune's orbit, which might have amounted to several astronomical units, would have a profound effect on the outer solar system. As Neptune moved outwards, its orbital period would have changed. As it did so, the positions of its mean motion resonances would also have moved. The resonances would have drifted outwards ahead of the planet and moved through the region occupied by the surviving outer solar system planetesimals. The idea of slowly moving resonances sweeping up planetesimals as they went along has been developed by another of the small band of theoretical astrophysicists trying to sort out the 4.5 billion year history of the solar system.

Renu Malhotra was born in New Delhi, India, in 1961. She grew up mostly in Hyderabad, a city more or less in the geometric centre of the Indian sub-continent. Looking back, she says that it seems she was fated to study physics. As a child she had many conversations about nature, the physical sciences and the atmosphere with her father, an aircraft engineer with Indian Airlines. These led her to read physics at the Indian Institute of Technology in Delhi. This is one of five major engineering campuses in India and one to which entrance is very competitive, with only about 1000 of the 200 000 potential students being admitted each year. After passing the nationwide entrance examination, and studying there for five years, Malhotra had gained what she describes as an excellent education.

From Delhi, she went to the United States. She entered graduate school at Cornell University at Ithaca, in New York State, to study for a PhD in physics. It was here, while working in the field of theoretical non-linear dynamics, that she discovered solar system dynamics and soon became interested in the problems it raised. 'After all', she says, 'The solar system is the pre-eminent dynamical system, all of modern physics came out of the study of planetary motions'. To Renu Malhotra the study of the dynamics of solar system seemed like a natural home for the things she loves to do, combining mathematical analysis and theoretical physics to understand the world around her. She gained her doctorate in 1988 then went to the California Institute of Technology in Pasadena for a period of postdoctoral research. She then moved to the Lunar and Planetary Institute in Houston, Texas.

Like Martin Duncan and Hal Levison, she spends her working time reading scientific papers, doing analytical studies of dynamical problems and running computer models on her desktop workstation. Inside the machine millions of years flow past in hours, as the primitive solar system evolves. Electronic planets grow and their orbits change as phantom collisions occur by the thousands. Eventually, some order emerges and a model of how the the solar system might appear is revealed, ready to be compared to real planets, moons and asteroids in our planetary system. Malhotra first used the idea of resonance sweeping to explain how Pluto might have been captured into its present orbit. She then expanded this idea to account for the trapping of other objects into planetary resonances. Her results suggest that this sweeping would be quite sufficient to clear out planetesimals from the region inside about 39 AU. As the orbit of Neptune expanded, objects would 'fall into' the resonances and would then be moved slowly outwards along with the planet. A consequence of this process, which can be tested by observations, is that not only must the regions swept clear be empty, but the resonances must contain more objects than could have got there by chance if the resonances had not migrated through the outer solar system, picking things up as they went along.

Although the resonance-sweeping hypothesis does explain the lack of objects inside 39 AU, and the large population of objects in the 3:2 resonance, it is not without its problems. For one thing, it predicts that the 2:1 Neptune resonance should also be heavily populated. If Neptune did migrate outwards, then its 2:1 mean motion resonance would have started out in a region containing large numbers of objects. These objects ought to have been picked up as the resonance moved outward. Indeed, if resonance sweeping was effective, then not only should the 2:1 resonance be quite full, but the area across which it swept should be quite empty. Unfortunately, this does not appear to be the case. Not only is the original location of the 2:1 resonance still occupied by large numbers of classical Kuiper Belt objects, but the present 2:1 resonance seems to be fairly empty. This may be because the objects there have escaped somehow, or because of some observational bias in the way objects have been discovered so far, or perhaps because the resonance sweeping did not occur exactly as the models postulate. In particular, the timescale of the resonance sweeping may be quite critical. The Japanese theorist Shigeru Ida has shown that the 2:1 Neptune resonance can be unpopulated if the orbital migration of Neptune happens in less than 5 million years or so.

Despite the difficulties with each model, and such problems are to be expected in a subject which is still in an early stage of development, it seems that a reasonable picture of how both the Plutinos and the classical Kuiper Belt beyond 42 AU can be sketched out. However, there remains another type of trans-Neptunian object to explain.

From where did 1996 TL$_{66}$, the object discovered by Jane Luu and others then assiduously recovered by Carl Hergenrother and Warren Offutt, come?

The existence of a population of objects on highly eccentric orbits which only graze the planetary system when they are at their closest to the Sun was first put forward by Julio Fernandez and Wing Ip when Fernandez was working in Germany during the early 1980s. More recent work, particularly numerical integrations of objects encountering Neptune done by Martin Duncan and Hal Levison in the late 1990s, showed that while most objects encountering Neptune had fairly short lifetimes, about 1% of the objects that they started with were still in orbits beyond Neptune after 4 billion years. It turned out that these objects could survive for such a long time because during their wanderings they were likely to spend quite a bit of time temporarily trapped in distant mean motion resonances with Neptune. This process can be illustrated through the medium of the numerical integrations by following one of these imaginary particles around the solar system. For 70 million years the object wandered randomly about near Neptune before it was trapped in Neptune's 3:13 mean motion resonance. It stayed here for 50 million years and then escaped, spending a little under 200 million years in limbo until it fell into the 4:7 mean motion resonance. It remained there for a further 340 million years. Leaving this resonance it then drifted into the 3:5 mean motion resonance where it stayed for about 500 million years. The test particle then escaped that resonance and wandered about the outer regions of the solar system for the remainder of the simulation.

If objects can survive this long, and if there were a significant number of objects in the Uranus–Neptune region soon after the solar system formed, it seems that there could be a large number of objects in this scattered disc today. This conclusion means that the first scattered disc object, 1996 TL$_{66}$, was not unique, it was just the harbinger of a much larger population of objects waiting to be discovered. More scattered disc objects have since been found but it may be a little while before this population can be mapped in detail. With orbits extending to 200 AU, most of the objects in the scattered disc will spend most of their time a great distance from the Sun. For much of the time they will be very faint. With present technology, only a few objects have yet been found when they were beyond 50 AU and an object at a 100 AU will be sixteen times fainter still. For the moment at least, only scattered disc objects that happen to be close to perihelion will be found by

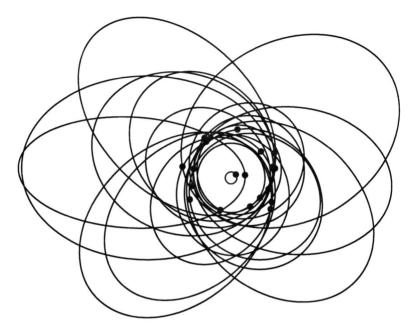

Figure 6.3 The orbits of the best observed scattered disc objects. The small circle in the centre is the orbit of Jupiter. The figure is a square with sides of 400 AU. (Chad Trujillo.)

current search programmes. The outer regions of the disc are likely to remain unknown territory for a while.

Before we leave this arcane world of phantom particles, chaos and resonances, it is worth taking a moment to review the state of our understanding this complex region. How does what has been learned relate to the problem of the origin of the short-period comets? Numerical integrations and analytical studies are in general agreement that the inner edge of the trans-Neptunian region occurs at about 35 AU from the Sun. Any object which finds itself inside this region cannot survive there for long before its orbit is drastically modified by the gravitational influences of the giant planets. The region from 35 to 42 AU is complex and the population here has been subjected to considerable dynamical evolution. Regions of stability exist around the mean-motion resonances. Objects librating around these locations avoid close approaches to Neptune and so can survive for the age of the solar system. Outside the mean-motion resonances, most orbits are not stable and these regions have long since been cleared of any primordial planetesimals. At the boundaries of the resonances are semi-stable states in which objects can survive for billions of years, but which are not protected forever. From time to time objects in these regions stray just a bit too far from safety and escape from the resonances. Once they do they may begin a journey towards the inner solar system and be destined to become a Jupiter family comet, or suffer a dramatic collision with one of the planets.

Beyond about 42 AU is a stable region where most orbits are fairly circular and of low inclination. This region contains objects that have probably been there for the age of the solar system and is the closest thing yet found to the disc of primordial material proposed by Edgeworth and Kuiper half a century ago. Ironically, it is fairly unlikely that objects from here will ever enter the inner solar system. Even though some of the orbits are chaotic, they tend not to evolve into paths which can encounter Neptune. Mixed in with these bodies is another population, the scattered disc. This comprises objects on more highly inclined and more eccentric orbits that dip inside the classical Kuiper Belt, but spend most of the rest of their time much further away, travelling a few hundred astronomical units from the Sun. These are objects which were originally formed in the Uranus–Neptune region and were subsequently ejected to great distances. However, orbits in the scattered disc are not stable over timescales of billions of years. Although they may spend significant amounts of time in outer mean motion resonances, scattered disc objects also have periods in which their orbits undergo chaotic wandering and which may bring them back to a close encounter with Neptune and perhaps a one-way ticket to the inner solar system. Beyond about 50–100 AU remains uncharted territory, where the gravitational effects of the planets are not significant, and to which we will return.

What are little planets made of?

CHAPTER 7

Wherever they come from, it has long been recognised that comets are very old and that they offer a way of studying the chemical composition of the early solar nebula. The recent history of the solar system is written in the atmospheres and surfaces of the planets and their moons, but most of the chemical evidence has long since been removed. More than four billion years of heating, asteroid impacts, geology and weather have altered the planets almost beyond recognition. Comets, on the other hand, have been preserved in deep freeze for billions of years and have probably experienced little thermal processing since they were formed. They are natural time capsules preserving a record of the long-vanished solar nebula. Many an application for telescope time has begun with words to the effect of, 'Comets represent the least processed solar system material available for study', but most of the people writing these applications, and perhaps even a few of those awarding them telescope time, must have known that this was not completely true.

The problem is that before a comet can be studied, it has first to be discovered. Comets are usually found by eagle-eyed observers who detect the faint hint of a diffuse coma around an otherwise starlike object. So, before a single detailed observation of it can be made, the comet has already started to change. The very thing which makes a comet bright enough to discover is the sublimation of the ancient ices which have survived on or near its surface since it was formed. Observations of comet Hale–Bopp, and of the Centaur Chiron, show that cometary activity is not restricted to objects close to the Sun. Comet Hale–Bopp had a distinct coma when it was more that 7 AU away and Chiron has been seen to outburst at even greater heliocentric distances. What this means is that active comets are not the least processed material in the solar system at all. That sobriquet surely belongs to trans-Neptunian objects, which have never felt the warmth of the Sun. It is this extreme age which makes observations of them so important. Unfortunately, the trans-Neptunian objects are quite small and very distant, a combination which makes them faint and difficult to study with all but the largest telescopes.

The first clues to the physical nature of the trans-Neptunian objects came during the early searches by Jane Luu and Dave Jewitt. The surveys were done using a filter which isolated mostly red light

Figure 7.1 The summit of Mauna Kea in Hawaii. Most of the physical studies of the trans-Neptunian population have been done from these observatories. At the extreme left is the tiny dome of a 0.6 m telescope, then the silver dome of the UK Infrared Telescope is partly hidden behind the University of Hawaii's 2.24 m telescope. The multi-national Gemini telescope has its dome slit open. Next to this is the 3.6 m Canada–France–Hawaii telescope. At the rear of the summit ridge are the Japanese Subaru telescope and the twin domes of the W.M. Keck observatory. The NASA Infrared Telescope Facility is the silver dome projecting from the square building centre right. In the valley are two submillimetre telescopes; the Caltech Submillimetre Observatory and the James Clerk Maxwell Telescope (JCMT). The building to the right of the JCMT is the maintenance facility for an array of small submillimetre telescopes which had not been installed when this photograph was taken. (Richard Wainscoat.)

since the combination of this filter and the response of their CCD gave the highest possible sensitivity. However, once some objects had been found, they were soon observed through other filters to determine their colours. The objective was to see where the trans-Neptunian objects fitted in the general scheme of minor planets, comets and so on. An obvious question was, 'Are the Kuiper Belt objects very red like Pholus, more-or-less neutral like Chiron, or are they some colour in between?'. The first observations were not conclusive. 1992 QB_1 had a fairly red colour, but 1993 FW seemed to be neutral.

This first hint that the colours of the Kuiper Belt objects might be diverse was soon confirmed. Luu and Jewitt reported that there was a wide range of colours amongst a dozen or so trans-Neptunian objects they had observed. A similar conclusion was reached by Kent University's Simon Green and other British observers who had been observing from La Palma. A European group led by Antonietta Barucci of the Meudon Observatory in Paris agreed. This was rather odd, since it suggested that the surface compositions of these supposedly primitive objects were not the same. Perhaps the surfaces of the trans-Neptunian objects were not ancient after all?

The conventional wisdom was that the Kuiper Belt would contain objects which were predominantly icy. They were expected to comprise a mixture of frozen gases such as carbon monoxide, ammonia and methane together with water ice and relatively small amounts of dust. However, while pure ices are blueish in colour, it is unlikely any ancient ices in the trans-Neptunian region would be pure. Their extreme ages would presumably have exposed them to at least some chemical modification. Although at such great distances there is not much solar radiation, indeed the Sun would appear little more than a bright star, Kuiper Belt objects would certainly have been bombarded by a steady flux of high-energy cosmic rays for billions of years.[†] Cosmic rays have sufficient energy that they are capable of breaking individual chemical bonds and when they encounter icy surfaces they smash their way a few metres into the material and do just that, breaking the bonds which hold together the simple molecules they encounter. The fragments of molecules which are produced then

---

[†] Cosmic rays are not rays at all, but charged particles of various types. They are mostly protons and electrons plus a few heavier ions which have been ejected from violent events elsewhere in the galaxy and are travelling through interstellar space at great speed.

recombine in unpredictable ways to form more complex compounds of larger molecular weight. The large organic molecules formed by this process tend to give the material a reddish hue, but the material does not stay red forever. Hydrogen atoms liberated during the process, being small, are able to escape from the solid material and do not always remain long enough to get chemically bound up again. So, if the bombardment continues, the icy surface becomes relatively rich in carbon. This excess of carbon, being black, tends to make the material dark. The mish-mash of chemical bonds formed also makes the material refractory, or hard to evaporate. The result is that a tough crust called an irradiation mantle is formed. No-one really knows how long it takes to make a refractory crust that is stable for long periods, but estimates tend to run in the range of 10–100 million years. Such a long timescale is impossible to duplicate in the laboratory, although it can be simulated by using very large doses of radiation over shorter periods, but it is quite short in terms of the age of the solar system. This suggests that all the Kuiper Belt objects ought to have refractory crusts and so be dark and uniform in colour. So, at first sight, a wide range of colours is rather hard to explain.

However, science is about trying to understand the unexpected and it was not long before several ideas were put forward to explain the apparent differences amongst these supposedly similar objects. The simplest suggestion was that the chemical composition of the objects was not the same to start with. This is by no means impossible; there are distinct trends in composition with heliocentric distance amongst other groups of solar system objects. For example, in the main asteroid belt between Mars and Jupiter, the innermost objects are more rocky and have less water bound up in them than those found further out. These trends are attributed to the relatively rapid fall in the temperature of the solar nebula at increasing distances from the Sun. The temperature gradient would have tended to drive volatile ices away from the inner edge of the asteroid belt, leaving behind dusty material which went on to form mostly rocky objects. However, this sort of mechanism is less likely to have been significant in the trans-Neptunian region. At these great distances from the Sun the range of temperature across the planetesimal formation region would have been fairly small, only 10 degrees or so. None the less, if objects in the trans-Neptunian region originally formed at a range of heliocentric distances before being transported outwards, by resonance sweeping or some other gravitational effect, then perhaps it is naive to believe that they all have the same initial composition. Adding to the mystery is that there is no agreement about trends in the colours with any orbital parameters. The different coloured objects all appear to be mixed together.

Another alternative explanation of the colour diversity is to invoke some kind of resurfacing which gradually changes the colours of the objects. This could happen if the surface was slowly developing a reddish colour under the influence of cosmic rays and solar radiation when something forced out fresh material of a different colour from inside. By analogy with the Centaur Chiron, which is fairly neutral in colour, cometary activity might be the cause of this. However, at such large distances from the Sun it is hard to understand from where the energy to sublime significant amounts of gas and cause a comet-like outburst might come. An alternative is that the resurfacing might be the result of collisions with other, smaller, trans-Neptunian objects. The impacts might punch through the dark crust to fresh material below it. Such an impact might produce a crater that not merely pene-trated to bluer material, but was surrounded by an ejecta blanket of fresh material from below the surface. Over time, which would proba-bly be measured in millions of years, this freshly exposed material would be irradiated and slowly darken. If this is what is happening today, then a range of colours is fairly easy to explain – the objects most recently subjected to a few large impacts will have the most exposed ices and so the bluest colours, the ones which have not been struck recently will be redder.

The impact resurfacing idea was advanced in the early 1990s, but has since been supported by work published by Susan Kern and her collaborators in October 2000. Observations of the Centaur 8405 Asbolus (1995 GO) made with the Hubble Space Telescope in 1998 suggest that Asbolus has regions of its surface covered with very dif-ferent materials. For technical reasons the observations of Asbolus were made as two pointings of the satellite with a gap of over an hour between them. Although the original intention had been to add the two datasets together, Kern found that the two observations looked rather different. It seemed that one region was uniformly dark and featureless while the other had a brighter spot, perhaps with evidence of water ice. This bright spot could be an impact crater which had bro-ken through the older crust and exposed fresh, icy material from below.

Of course, this result was unknown in 1998 and further observations tended to cloud, rather than resolve the issue. The first hint of a problem came with a short but possibly important contribution from Steve Tegler and Bill Romanishin. Tegler, like many other astronomers of his generation, had become interested in astronomy as he followed the exploits of the Apollo astronauts exploring the Moon in the late 1960s and early 1970s. His first encounter with comets came in 1973, at the age of eleven. Like many other amateur astronomers of the time he tried and failed to find comet Kohoutek, which turned out to be much fainter and harder to locate than had been predicted. He had more success with Comet Halley, studying it with telescopes in Chile as part of a postgraduate programme which eventually led to a PhD from Arizona State University in 1989. Whilst in Arizona he had been introduced to the problems of CCD photometry by Bill Romanishin, then a postdoctoral researcher interested in studying faint galaxies. In 1995 Tegler moved to Northern Arizona University in Flagstaff where he once again encountered Bill Romanishin. His old observing partner was now on the staff of the University of Oklahoma and was visiting Arizona for the summer. Since they had a lot of experience of photometry of faint objects, they decided to try their luck with observations of some newly discovered trans-Neptunians. They were able to get observing time on the Kitt Peak 2.3 m telescope and had their first run in November 1995. Tegler describes their early observing runs as 'A struggle'. Sometimes the objects' orbits were so uncertain that they could not be found in the relatively small field of view of their CCD camera. Sometimes the objects were much fainter than predicted and so were impossible to measure accurately.

Despite these difficulties Tegler and Romanishin pressed on with their programme. In 1998 they published a paper in the journal *Nature* in which they suggested that the Kuiper Belt objects, and their cousins the Centaurs, were divided into two distinct classes of basically similar objects. Their results seemed to show that one class was neutral in colour, or Chiron like, and the the other class was very red, rather like Pholus. Such a bimodal distribution of colours would imply that there were compositional differences between the two classes, perhaps relating to the region of the protoplanetary nebula in which they formed. Significantly, a bimodal distribution is very difficult to explain in terms of gradual resurfacing which clearly predicts that there should be a wide range of colours and not two narrowly defined classes.

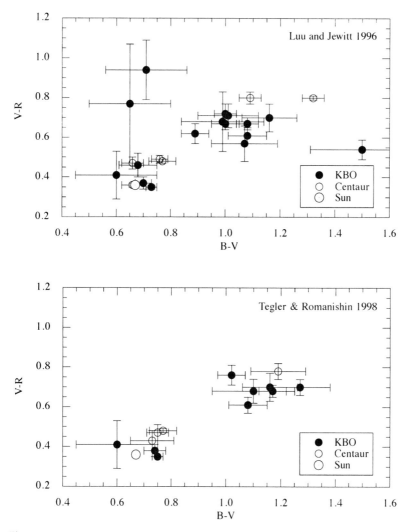

Figure 7.2 Two BVR colour diagrams for trans-Neptunian objects and Centaurs as presented by different groups. In (a) Jane Luu and Dave Jewitt's data suggest that the objects have a wide range of colours. (b) shows observations by Steve Tegler and Bill Romanishin which indicate that there are two distinct groups of objects with either neutral (bottom left) or red (top right) colours. (Dave Jewitt.)

Several other groups doing similar projects did not agree with the conclusions of Tegler and Romanishin, but worse was to come. They could not seem to agree amongst themselves either. Part of the problem may have been that there were few observations and the various groups often tried to bring together data from a number of different sources. Since each observer was using different telescopes, and perhaps subtly different filter systems, this lack of consistency might have produced more scatter in the data than anyone realised at the time. Dave Tholen, who tries to do photometry with such precision that he is sometimes known as 'Dr Milli-mag', speculates that part of the problem might have been due to faint background galaxies. There are faint galaxies all over they sky and, because they are extended, they are quite hard to spot. However, even though they are faint, the total light from a distant galaxy may be significant over an area of a few square arcseconds. So if a faint trans-Neptunian object happens to be in front of a galaxy when it is observed, considerable galaxy light may be included in the measurement unintentionally. The result will be a value which is too bright. Repeating the measurement a few hours later when the object has moved onto a different patch of sky, which may or may not have a different galaxy in it, will produce a different result. The lack of agreement between different groups was a worry, and deep down was the hint of something that few photometrists like to admit. Photometry is supposed to be a science, but sometimes it seems to resemble a black art. Photometry of moving 23rd magnitude objects from moderate-sized telescopes presents a formidable observational challenge which can be approached in a number of different ways. Perhaps some of the techniques being used were not as good as was first thought.

At the root of the debate was how to extract the most accurate value for the flux of a faint Kuiper Belt object from a CCD image. A few years ago the problem was much simpler. Most photometers used single detectors, such as photomultiplier tubes, that viewed the sky through a fixed aperture. The detectors just counted the number of photons which arrived in a given time. The data that was output was, in effect, a single number per observation and that was that. The field of view was usually defined by a physical aperture made by drilling or etching a hole through a piece of metal or metallic film and putting this into the beam of light somewhere between the telescope and the detector. The aperture defined a circle on the sky and admitted light from only that region. When observing, the effects of seeing and errors in the telescope tracking spreads out the starlight, but provided the aperture is fairly large, the detector sees almost all of this light. This of course is good, but the large aperture means that the detector also sees considerable background sky around the target. This large background is bad because even on a dark night many photons arrive from the sky and random fluctuation in the rate at which they arrive constitutes a source of noise in the measurement. The more sky there is in the aperture, the larger the error in the photometry produced by the sky noise. (The error is proportional to the square root of the number of photons which are arriving.) Sky noise can easily swamp the signal from very faint objects and this argues against using large apertures for photometry of faint objects. Switching to a smaller aperture reduces the sky noise by allowing through less background light, but it also means that it becomes more important to ensure that the object being measured is accurately centred in the photometer's aperture. If the target is not well centred, then some light from the target is lost. More to the point, the amount of light which is lost varies from object to object depending on how well each target is centred and on the seeing at the time of the measurement. Compromises had to be made to try and get the best ratio of signal to noise, and typically astronomers doing photoelectric photometry used apertures which projected 10–20 arcseconds on the sky.

The advent of cameras fitted with CCDs and infrared arrays revolutionised photometry. Their greater sensitivity made it possible to measure much fainter objects than before, but it also provided several challenges in deciding how to measure them. With a traditional photoelectric photometer, the observer made a choice of what aperture to use, put the aperture slide into the beam and faced the consequences. If, later on, it was realised that the original choice was not the best one, it was just too bad. It was already too late to do anything about it. CCD cameras opened up a huge number of new variables since each frame recorded images containing many data points, first thousands and soon, as the chips got bigger, millions. Before any photometry could be done, these images all had to be processed. Different observers like different processing methods and use different software packages to remove bad pixels and flat field their images. While the various methods should all produce the same result, great care has to be taken to ensure that, for example, the flat-field is really flat.

However, the real trouble usually starts when it comes to actually doing the photometry. By analogy with traditional techniques, a favoured method is to use image processing software to draw a circle around the target and then to add up the signal in each of the encircled pixels. This is analogous to measuring all the flux through a single aperture. Of course the pixels are usually square, and circles are round, so at the circumference of the software apertures some allowances have to be made where the circles cut across the square pixels. Just as in classical photoelectric photometry, the smaller the software aperture which can be used the better, but the smaller the circular aperture becomes, the larger is the fraction of encircled pixels which fall under the circumference of the circle. So as the aperture gets smaller the software's compensation for these edge effects becomes more critical. Not only that, but since the astronomer chooses the aperture after the observation rather than before, various different apertures can be used on the same image. These different apertures don't generally produce the same answer. This is because the light from the objects under study is spread out by the telescope optics and the seeing into a little blob on the image. Different software apertures will include different amounts of light from the object and background sky. Unless the images are very sharp, small apertures tend to miss quite a bit of the light that the effects of seeing and errors in focus have allowed to fall outside the aperture. Since the calibration of the observations will have been made by observing stars in different patches of sky at different times (when depending on the seeing the star images might have been sharper, or more fuzzy, and so more or less light might have been concentrated in the central regions of each stellar image), care must be taken to allow for these fluctuations.

A technique commonly used when measuring faint targets is to use very small apertures. This excludes most of the sky and so maximises the signal-to-noise ratio. However, this gain in signal to noise comes at the expense of losing some of the light, and getting a value which in absolute terms is too small. To correct for this, the amount of light lost from the small aperture is then estimated by observing a much brighter star in a series of different apertures. Since the bright star produces images with high signal to noise in even very large apertures, it is possible to trace out the amount of missing light in progressively larger apertures and calculate what is called an aperture correction. This correction can then be applied to values from the small apertures to allow for the missing light. Aperture correction is another idea that sounds simple, but is actually much more difficult than it appears. A variety of problems lurks to trap the unwary. These include, but are not limited to, picking a reference star that is in fact not a star at all, but is a distant galaxy and so does not have the correct profile to start with, and having the CCD chip not exactly at right angles to the incoming beam so the focus, and hence the aperture correction, is different at different places in the image. If this was not enough, moving objects smear out during a long exposure so the image of the asteroid will not have the same profile as a star in the first place.

Romanishin and Tegler attacked this problem very carefully. They kept their exposures short to reduce the effects of image smearing. They added together the same images in both the fixed reference frame of the stars, to get the best possible aperture corrections, and then in the moving reference frame of the asteroids so they could safely use very small apertures. The small apertures boosted the signal-to-noise ratio and minimised the likelihood of faint galaxies contaminating their measurements. Other groups abandoned apertures altogether and used software which fitted the observed profile of the objects in the images to a shape predicted from statistics. It then calculated the total flux using this profile. Although well suited to faint stars, this method also runs foul of the fact that an asteroid will move during the exposure and so its profile is not exactly what the software is expecting.

Ultimately, there is no getting away from the fact that this is a very tricky problem and it is hard to know which method produces the best results. Each approach has its champions who often defend their results with conviction. Even so, in the back of everyone's mind is the fact that there is only one right answer and sooner or later the truth is going to come out. Fortunately, most of the individuals involved in this debate recognise this problem and have been cooperating to try and reach a solution. At a scientific workshop held in Germany in November 1998[†] it was agreed to hold a blind test. A number of people would try different methods on the same images and see how closely the answers came out. Catherine (known to her friends as Cathy) Delahodde of the European Southern Observatory in Chile agreed to organise this test. She took some CCD frames containing a genuine trans-Neptunian object and added 24 synthetic objects for which only she knew the correct magnitudes. These frames were posted on an internet site and various solar system astronomers were invited to download them and try to reduce the data using their favourite methods. The odd thing was, all the results came out almost the same, with most of the groups getting values that agreed with each other to within a few per cent. What was going on? Was it possible that the reasons for the disagreements were not due to different techniques at all? The observations were not all made at the same time so perhaps

---

[†] Actually in a hotel bar after the end of the day's official presentations.

the objects themselves were varying, either due to rotation or to some kind of comet-like activity. Both such mechanisms have been proposed, and both are worth considering.

As noted in connection with the Centaurs, repeated photometry of a solar system object may reveal that its brightness varies over time in a manner that is not explained by changes in its distance from the Earth and the Sun. If the changes are irregular, as was the case for Chiron in the 1980s, then it is most probably due to a cometary outburst. Such outbursts can surround the object with a cloud of gas and dust which reflects sunlight and makes it appear brighter. Although in the case of comets fairly close to the Sun the coma is usually visible quite easily, faint comae around very distant objects may be impossible to resolve directly. An alternate approach is to compare the image profile of the suspected comet to the profile of a star of similar brightness; if the object is a comet its image will seem slightly wider and less pointlike than a star. Given that objects beyond Neptune are generally very faint, and any extension is likely to be small, this sort of comparison is very difficult to do from the ground. A particular problem is the effect of seeing, which blurs out the profile of each object and may mask any faint extension close to the suspected comet. To get around this difficulty there have been attempts to use the Hubble Space Telescope to detect comae around Kuiper Belt objects. The Hubble Space Telescope has very fine optics and is unaffected by atmospheric turbulence, so it can produce very sharp images, well suited to image profile comparisons. Although one group announced in 1988 that they thought they had found evidence that the profile of one Kuiper Belt object seemed to be extended, this result was not confirmed by more careful analysis. Direct evidence for cometary outbursts in the Kuiper Belt thus remains lacking.

Outbursts produce irregular and unpredictable changes in brightness, but rotation can cause a solar system object to wax and wane on a regular and repeatable basis. The variation may have one of two root causes. A spherical object which is the same colour all over, for example a billiard cue ball, will appear the same brightness whichever side is facing the observer, but a rotating object which has large regions of different reflectivity will vary in a regular way. At some times the observer will see entirely the brighter side, sometimes a bit of both and sometimes the darker face will rotate into view. This effect will produce a regular lightcurve with one maximum and one minimum per rotation. Of course this is only true if the different regions are large and cover appreciable fractions of a hemisphere. If the different regions are evenly distributed, like a soccer ball with black spots spread uniformly over a white surface, the average brightness will not change during a rotation because the different areas will blur together and appear a uniform grey.

If the object is irregular in shape, like a potato, then as it turns it will present different faces to the observer. Since the projected area will change, the apparent brightness will go up when the wide face swings into view and down again when the object is seen end on. In this situation, the lightcurve will show a characteristic double peaked lightcurve with two maxima and two minima (which need not be the same) per rotation. The range of variation of such a lightcurve gives an idea of how irregular the object is. Generally speaking, the bigger the change the more irregular the object, although this effect varies depending on the viewing geometry. If the object is seen close to pole on it will have almost no lightcurve, no matter what shape it is.

The first attempt to produce a lightcurve of a trans-Neptunian object came from the discoverers of 1993 SC, Iwan Williams, Alan Fitzsimmons and Donal O'Ceallaigh. Their inadvertent re-observation of this object during the search programme described earlier produced a total of 11 observations of 1993 SC over a period of six nights. As the object moved slowly, it remained in more-or-less the same starfield over the whole observing run. This made it possible to compare its brightness with the same stars on every night of the run. These comparisons seemed to indicate that 1993 SC was varying by more than half a magnitude, which in turn suggested that it was quite irregular in shape. Williams and his team used this lightcurve to estimate a rotation period of about 15 hours for the object, although they did admit that with such sparse data the result was not likely to be very reliable. Just how unreliable it was became clear a couple of years later when two different groups re-observed 1993 SC and found almost no variation at all. It now seems that Williams' group was misled by unsuspected errors in the photometry extracted from their images. In retrospect, perhaps these problems should have sounded a warning that photometry of these objects was harder than it seemed.

Despite the difficulties of trying to measure lightcurves, a few people have persisted in the effort. First to publish some detailed results were Bill Romanishin and Steve Tegler, who used the same dataset on which they had based their conclusions about the colours of distant asteroids. They found that many of these objects had no detectable lightcurves and in several cases the maximum possible variations were only a few per cent, the likely precision of their measurements. Unlike similar studies of brighter objects, for which it is possible to build up a lightcurve containing many points during a single rotation, the data for the Kuiper Belt objects was generally rather sparse. Even so, in a few cases, it did suggest that there were significant variations over periods of a few hours. If these changes were due to the rotation of a spherical object with hemispheres of very different reflectivities, the resulting single peaked lightcurve would imply rotation periods of only a few hours. This is so rapid that the objects would be hard put to stay together; the rotation speed at the equator would be rapid enough to throw material off the surface and into space. Accordingly, Romanishin and Tegler interpreted the data as double peaked lightcurves from irregularly shaped objects. This gave rotation periods of 6–10 hours, similar to main belt asteroids in the same size range, and more physically realistic.

Romanishin and Tegler took their conclusions one step further. They noted that the objects with measurable lightcurves were the intrinsically faintest ones and that the objects which did not vary were generally brighter. They suggested that this was because the larger objects are massive enough that their gravity causes them to collapse into a sphere while the fainter, and so smaller ones, cannot collapse and so are irregular in shape. Based on an assumed reflectivity of 4% for a typical Kuiper Belt object, they estimated that the transition size from spherical to irregular was a diameter of about 250 km.

Other groups have tried to follow up this work, but it remains very difficult. Alan Fitzsimmons' group observed two objects, 1996 TO_{66} and 1994 VK_8 in 1997. Although they did detect what seemed to have been significant changes in the brightness of 1994 VK_8, they were unable to confirm any particular rotation period from their data. They found no detectable lightcurve for 1996 TO_{66}. A rather more interesting conclusion about 1996 TO_{66} came from a European team. Olivier Hainaut, Cathy Delahodde, Antonietta Barucci and Elisabetta Dotto had determined a lightcurve for 1996 TO_{66} based on observations taken in 1997 and 1998 from Chile. These observations showed an almost symmetrical double peaked lightcurve indicating a shape-dominated rotation period of 6.25 hours. The peak-to-peak amplitude variation during the rotation was found to be 0.12 magnitudes. Just as they were preparing their result for publication, they learned that 1996 TO_{66} had been observed in September 1998 from Mauna Kea by Karen Meech and

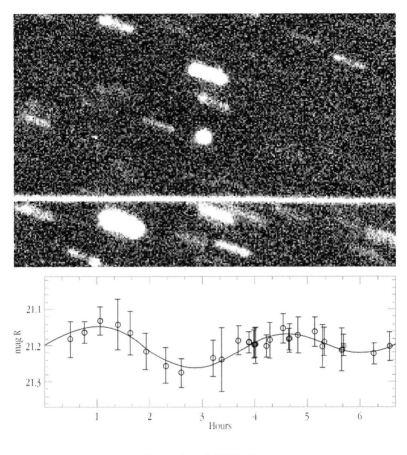

Rotation Period of 1996 TO66

©European Southern Observatory

Figure 7.3 Top. An image of trans-Neptunian object 1996 TO$_{66}$. Bottom. The light curve deduced from observations made at the European Southern Observatory. (ESO.)

James 'Gerbs' Bauer. Olivier Hainaut had been a research fellow in Hawaii before moving to Chile and had worked with Karen Meech before, so they agreed to combine the data from all three runs. Interestingly, the new observations were not consistent with the lightcurve derived from the two previous years. The 1998 observations suggested that 1996 TO$_{66}$ had a single peaked lightcurve with a much larger amplitude of 0.33 magnitudes, although the rotation period seemed to be exactly the same.

Hainaut and his co-workers examined several possible explanations for this and concluded that the best one was a brief phase of cometary activity occurring sometime between the 1997 and 1998

129

observations. They suggested that during the outburst a region of the object's surface was modified, probably by the deposition of fresh bright ices which covered up some otherwise darker material. Such activity might, indeed probably would, also change the observed colour of the object, perhaps explaining some of the colour differences reported by various groups. Further observations of the same object were made in 1999 using the 8.2 m European Very Large Telescope in Chile. This data implied a similar rotation period, but the amplitude of the lightcurve had changed again. This time the variation was about 0.2 magnitudes and the shape of the curve was different. According to Hainaut, this could confirm the hypothesis that the object is being actively resurfaced on short timescales.

For the moment, many of these questions about lightcurves and surface colours remain unresolved. The observations are difficult, the signal-to-noise ratio is often low and the uncertainties in the data may be larger than people are yet willing to admit. It will be a little while before we know the details for certain, but what is already clear is that the physical properties of the objects in the Kuiper Belt are at least as complicated as the orbital dynamics. Whatever the final resolution of the debate over the colours and rotation rates, photometry is unlikely to solve the problems of the chemical composition or mineralogy of the outer solar system. Two-colour diagrams are useful for sorting objects into broad groups, which may have roughly similar surfaces, but the filter passbands cover a wide range of wavelengths. Filter photometry may mask a variety of spectral features which could be used to identify specific minerals, or molecules, if only the objects could be observed with much higher spectral resolution.

Although Dave Jewitt and Jane Luu tried to obtain an optical spectrum of the relatively bright Plutino 1993 SC as early as 1994, spectroscopy of Kuiper Belt objects, especially in the near infrared region where ices have identifiable spectral features, is not really a practical proposition with 2–4 m class telescopes. It was not until solar system astronomers got their hands on bigger mirrors that much progress could be made in understanding the composition of these distant objects. The first results had to wait for spectroscopy from the huge Keck telescopes on Mauna Kea.

The W. M. Keck Observatory comprises two of the largest optical telescopes in the world and a headquarters building in the sleepy cattle town of Waimea, on the Big Island of Hawaii. The observatory was financed by a donation from the W.M. Keck Foundation to a consortium comprising the California Institute of Technology and the University of California. The objective was to build a new large telescope to take advantage of the clear skies available over Mauna Kea. Since, at the time the project was conceived, making very large single mirrors was known to be very difficult, the Keck telescopes are of an innovative design. They comprise numerous hexagonal mirrors, each about 2 m in diameter, fitted together to make a single collecting area equal to a mirror 10 m across. The whole array of hexagonal mirrors is supported by an ingenious backing structure which keeps them all pointing in the same direction so they function as a single mirror. So successful was this concept that once the first telescope was operational, the Keck foundation was persuaded to provide funds to build a second telescope, increasing the total sum they had donated to about $140 million.

Since the Keck Observatory was owned by the Californian consortium, all the observing time, apart from the fraction claimed by the IfA in Honolulu for allowing the telescopes to be built on Mauna Kea, originally belonged to these two establishments. However, from October 1996 NASA entered into an agreement to provide some of the running costs of the Keck Observatory in return for one third of the observing time on one of the two telescopes. NASA, which already funded the much smaller IRTF telescope on Mauna Kea, needed access to a large telescope for its wide-ranging programme for studying the origins of planetary systems. Contributing to the Keck Observatory was the quickest and most cost-effective way of doing this.

In early October 1996, some of this observing time was allocated to a group including the University of Arizona's Robert (Bob) Brown and NASA's Dale Cruikshank. They used it to take infrared spectra of the Plutino 1993 SC with the Keck observatory's dual-purpose near-infrared instrument NIRC. Although NIRC stands for Near InfraRed Camera, the instrument can also be used as a low resolution spectrograph. Spectroscopy is done by inserting some extra optics into the camera's optical path and using a grism (a combination of grating and prism) to disperse the incoming light into a spectrum. For this to work, the grism mode includes a narrow slit which is projected onto the sky and which blocks most of the field of view. Only light passing though the slit reaches the grism and is dispersed into a spectrum. Accordingly, the image of the object under observation has to be placed exactly in the slit and kept there throughout the observation. Luckily, the fact that NIRC also works as a camera makes this exercise of finding a faint target and getting it lined up with the slit slightly easier than it would be for a pure spectrograph. It is possible to switch NIRC between the imaging and spectroscopy modes to see exactly where the target appears in the instrument's field of view. Brown and colleagues started their observations by pointing at the predicted position of 1993 SC and taking a couple of images using the instrument in its camera mode. By comparing these two pictures, they quickly identified which object was their target and moved the telescope to place 1993 SC in the correct place to send its light through the slit. Once this was done they switched to spectroscopy mode. As is usual with spectroscopic observations, they also observed some stars in order to allow for the effects of the Earth's atmosphere, which does not transmit equally well at all wavelengths. After calibrating their data and using data from their comparison stars to remove any features due to the atmosphere, they had the first infrared spectrum of a Kuiper Belt object. What did it mean?

The spectrum was fairly noisy, which is not surprising since their target was so faint, so they smoothed it by combining some of the points to improve the signal-to-noise ratio at the expense of lowering the spectral resolution. The result was an interestingly complex spectrum that seemed to show a number of spectral features. Although the resolution of the spectrum was not good enough to identify the features unambiguously, the overall shape was rather similar to spectra of Pluto and of Neptune's largest moon Triton. Spectra of Pluto and Triton taken during the 1980s suggested that their surfaces might include some methane ice trapped in a solid solution of frozen nitrogen, and while the spectrum of 1993 SC did not prove that the exact same situation existed there, it was certainly suggestive of a surface containing a frozen mixture of light hydrocarbons. The existence of solid nitrogen on 1993 SC was a bit more problematical, since given the likely temperature of the surface (about 50 K, 50 degrees above absolute zero or -223 degrees Celsius) and its low gravity, it is not certain that solid nitrogen could survive there over the age of the solar system.

A second spectrum of a trans-Neptunian object was not long in coming. Dave Jewitt and Jane Luu observed the scattered disc object 1996 TL_{66} from two different telescopes in September 1997. They used the NIRC camera/spectrometer on one of the Keck telescopes to take an infrared spectrum and combined this with an optical spectrum taken at the Multiple Mirror Telescope, or MMT, located on Mt

Graham in Arizona. The MMT is another telescope of unusual design. In its original form it combined the light from five mirrors at a single focus to produce the light-gathering power of a single 4.5 m mirror. However, unlike the Keck telescopes, these mirrors were not hexagons tiled together into a single surface, but conventional circular mirrors mounted in a common framework. The design worked, but the telescope was closed in 1998 to remove the complex multi-mirror system and replace it with a single mirror 6.5 m in diameter.[†]

Jewitt and Luu finished up with three spectra, one from the MMT covering the optical region and two from the Keck telescope. Between them these three spectra encompassed the entire optical and near-infrared region. They carefully combined the different spectra, using photometric measurements taken in both the optical and infrared regions to ensure that they did this correctly. The final result was completely featureless. The spectrum showed no sign of the infrared features which Bob Brown and his group had found in 1993 SC. There was no evidence for water ice or any hydrocarbons. All in all, the spectrum of 1996 TL_{66} was rather dull and featureless, just like the spectra of some of the Centaurs.

A third spectrum of a trans-Neptunian object was published in 1999. Once again, this came from Bob Brown and his collaborators using NIRC on one of the Keck telescopes. This time the target was 1996 TO_{66} and the result resembled neither 1993 SC nor 1996 TL_{66}. The spectra, which were taken and reduced in a similar manner to their earlier data, showed two absorptions in the 1–2 micron region. These features suggested that there was water ice on the surface of 1996 TO_{66}. Not only this, but spectra taken on two successive nights showed that the amount of water, as determined by the depth of absorption features in the spectra, was variable. Here, at last, was evidence that the Kuiper Belt objects might have global variations in their surface properties. Perhaps the surface of 1996 TO_{66} was patchy, with some regions containing more fresh ice than others, as suggested by the changing lightcurve reported by Olivier Hainaut and his colleagues.

The likely presence of ice on 1996 TO_{66} was confirmed by observations made in 1998. During August and September, Keith Noll, Jane Luu and Diane Gilmore attempted to observe five trans-Neptunian objects using the NICMOS near-infrared photometer/spectrometer

---

[†] Despite this change, there are no plans to rename the telescope, it will still be known as the MMT.

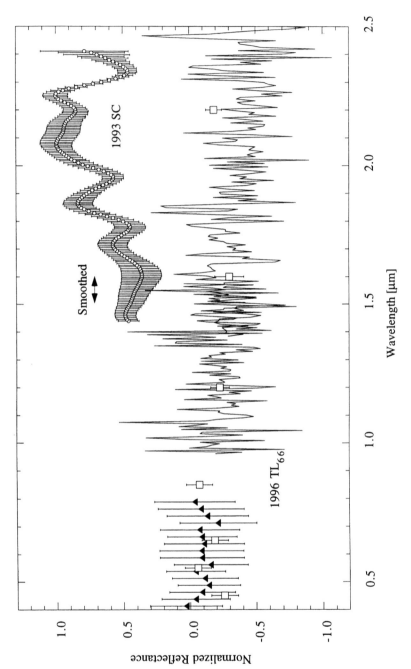

Figure 7.4 Spectra of two trans-Neptunian objects. At the top is the spectrum of 1993 SC taken by Bob Brown at the Keck Telescope. Below this is

on the Hubble Space Telescope. The objects were too faint for spectroscopy, but in each case they planned to take measurements through a series of six filters covering the 1–2 micron range. Combining these would give a rough spectral fingerprint of each object and should reveal any significant spectral features. One observation missed its target. The satellite had been pointed at the position requested, but the object was not to be seen. This was because the best orbital elements then available for the object were not correct, causing the telescope to be pointed at the wrong position. The other four observations all went well. After applying the usual flat field corrections and removing cosmic rays from the images, Noll and his co-workers extracted the best photometry they could from their data. They then combined their infrared data with some optical observations from the literature and looked at the results. Their data showed that 1996 TO_{66} was rather different from the other three objects which they had observed. 1996 TP_{66}, 1996 TQ_{66} and 1996 TS_{66} all got redder at longer wavelengths and had a few dips that hinted at some as yet unknown spectral features. In some ways these three objects were broadly similar to the redder Centaurs. On the other hand, 1996 TO_{66} was bluish and the photometry through the various infrared filters showed that it had dips in its spectrum. These dips were consistent with Bob Brown's 1999 Keck spectrum showing water ice on the surface.

The presence of ice on 1996 TO_{66} was further confirmed by Mike Brown, of the California Institute of Technology. As a member of the Caltech faculty Brown has access to the Keck telescopes and he had used them to survey six Kuiper Belt objects. However, no ice was seen in the spectra of any of the other five objects. All appeared featureless, which Brown says he found interesting. Like many other people he had expected the objects to have spectral features typical of those seen on the red Centaur Pholus. After all, if Pholus looks the way it does because it has only recently left the Kuiper Belt, then the objects still there ought to look just like it. The fact they do not casts doubt on the commonly accepted view that Pholus has a very old surface.

Mike Brown finds his results strange, and wonders if the Classical Kuiper Belt might contain objects from two distinct populations. However, Noll, Luu and Gilmore sounded a warning that this dichotomy did not, in itself, support the case for a bimodal distribution of trans-Neptunian objects. While it might appear red colours would correspond to objects dominated by organic rich solids, and blue ones to objects dominated by water ice, the situation was clearly more complex. The red Centaur Pholus has water ice features in its spectrum, but the bluish trans-Neptunian 1996 TL_{66} does not show any evidence for ice.

Supporting what might be described as this 'conclusion of confusion' are a number of ground-based observations. Dave Jewitt and Jane Luu observed five objects in the infrared JHK bands using one of the Keck Telescopes. They combined their data with visible photometry taken from the University of Hawaii 2.24 m telescope a few weeks later to determine visible–infrared colours for five objects. Soon after, John Davies, Simon Green, Neil McBride, Dave Tholen and their students went one step further. They used the University of Hawaii 2.24 m and the nearby UKIRT telescopes to observe a dozen or so objects in the visible and near infrared simultaneously. By observing in this way, it was possible to remove any uncertainties in the colours which might result from the unknown effects of lightcurves or hypothetical cometary outbursts. This observing strategy involved much telephoning from dome to dome throughout the night in order to synchronise the observations. Both groups obtained sets of visible–infrared data which indicated that there was a wide range of colours that were not correlated with anything in particular.

However, Steve Tegler and Bill Romanishin soon returned to the fray with a new result. The controversy surrounding their initial announcement that the trans-Neptunian objects fell into two well-defined colour groupings had enabled them to get time on the Keck telescopes to continue their programme. In 1998 and 1999 they observed 17 more objects from the Keck telescopes and in August 2000 published a paper in *Nature* defending their original conclusions. They also noted that there seemed to be a correlation of colour with distance from the Sun. They reported that objects in near-circular orbits beyond about 40 AU from the Sun were consistently very red, while objects closer in could fall into either of their neutral or red classes.

Tegler and Romanishin's new data were presented in the form of a talk at the October 2000 meeting of the American Astronomical Society's Division for Planetary Sciences in Pasadena, California. Immediately after it was finished, Dave Jewitt rose to present his results. Jewitt said firmly that his data, which also included observations from the Keck telescope, showed neither a bimodal colour distribution nor any trends of colours with distance from the Sun. Perhaps as the organisers had intended, it was in direct contradiction of the talk which had just finished.

There was no resolution of the issue at the meeting. Later one of the audience remarked that the bimodal distribution seemed to get less convincing as more data was presented, which is the opposite of what would happen if the effect was real, but Steve Tegler says he is sticking to his conclusions. Antonietta Barucci and her colleagues have continued their observing programme and are in the other camp; like Dave Jewitt they believe that their data do not support a bimodal distribution nor any trends with distance from the Sun. The issue is clearly a complicated and controversial one, and it is going to take some time and effort to unravel it.

Numbers
and sizes

To take full advantage of what the trans-Neptunian region can reveal about the formation of the solar system, astronomers need to understand more than just the physical characteristics of a few dozen objects. They also require details of the size distribution of the population to compare with their models of how the planets formed. How much material is there beyond Neptune? Is this consistent with observations of other stars thought to be forming planets?

Size distributions are generally such that there are many small objects for each large one. They are usually expressed in terms of the number of objects larger than a certain size plotted against that size. Since the number of objects increases rapidly as the size goes down, and the graph must represent a wide range of sizes from a few near-planet-sized objects to an uncountable number of dust grains, both axes are normally plotted on logarithmic scales. The result is usually a more-or-less straight line whose slope is called the power law index. The power law index for a population which contains objects which are still growing is different from that of a population which has stopped evolving, or one which is in the process of grinding itself back down into smaller pieces. So, the size distribution is critical to understanding the evolutionary history of the trans-Neptunian region.

Determining the size distribution of objects in a laboratory is just a matter of measuring and counting the sample. Attempting to do the same thing for a population of faint astronomical objects, most of which have not even been discovered yet, is rather more problematical. One of the difficulties is the issue of selection effects. Just how representative of the population as a whole are the objects discovered so far? While there is no need to discover every last object before trying to determine a reliable size distribution, it is important to make sure that the sample which has been observed is a fair selection of those that remain to be found. Samples of astronomical objects are notoriously loaded with observational biases and many people have found that failure to take these effects into account in statistical studies can result in seriously flawed conclusions.

As an example of an observational bias, consider that before they can be included in any census of the trans-Neptunian region, the objects must be discovered. Broadly speaking, the likelihood of a solar system body being discovered depends on its brightness. The brighter it is the more likely it is to be found using a medium-sized telescope, and there are a lot more medium-sized telescopes around the world than there are really big ones. At any given time the brightness of a solar system object depends on its size, its reflectivity and, since it shines by reflected light, on its distance from both the Earth and the Sun.[†] Since the trans-Neptunian region begins at about 40 AU, and the Earth's orbit is by definition at 1 AU, the precise observing geometry does not make much difference to the distances involved, but it is clear that for a given size and reflectivity, the further an object is from the Sun the fainter it will be. The fainter it is, the harder it is going to be to discover. This effect is seen very easily in the statistics of the known trans-Neptunian objects. About one third of them are Plutinos, but this does not necessarily mean that the Plutinos are this common. It just shows that if all other things are equal Plutinos, being on average closer to the Sun, are easier to discover.

There is another more subtle factor which mitigates in favour of the Plutinos. To make it into the list of known trans-Neptunian objects, an object has to be not just discovered, but re-observed on subsequent months. Only then can its orbit be determined. Since the observed arcs of most newly discovered distant objects are small, there is often not enough information to do much more than make a good guess at some of the orbital elements. Consequently, projecting an object's position forward so it can be reobserved a few weeks or months later often requires the staff of the Minor Planet Center to make a few assumptions. One of the assumptions they may make is that the object is in an orbit stabilised by a resonance, such as the 2:3 mean motion resonance with Neptune which marks an object as a Plutino. If this assumption is correct, then the object will duly appear

[†] There are a number of other subtle factors which affect the observed brightness of solid, rough, solar system objects and which depend on the exact angle between the Sun, the object and the Earth, but these do not concern us here.

close to the prediction, be recovered and added to the catalogue of known minor planets. However, if this initial guess is wrong, then the calculated future positions may be in considerable error and the object may not be where the observers trying to recover it are looking. If this happens then it is likely that the object will not be recovered. Unless it is discovered again by chance, it will join the legion of lost trans-Neptunians. What may have been an early example of this was the fate of 1993 RP, one of the first trans-Neptunian objects discovered. Brian Marsden assumed that, like the other objects discovered that autumn, it was in the 2:3 resonance. He made his predictions accordingly, but 1993 RP was never recovered. Perhaps it was not a Plutino, but something else, for example a scattered disc object. Unless 1993 RP is found again by chance, we will never know. The result of this sort of bias is that a higher fraction of Plutinos and other objects in orbits stabilised by likely resonances are recovered and they swell the list of known objects in a disproportionate way.

Additional observational biases can arise from the way in which the sky is searched. We saw earlier that during their long-running search programme, Dave Jewitt and Jane Luu concentrated on regions of the sky where the ecliptic plane was well separated from the Milky Way. They, and others, did this to increase their chances of finding faint moving objects by restricting the number of stars in each of their images. However, the use of such observing strategies means that some regions of the sky have not been searched, or have been searched with a lower probability of success, than others. If the trans-Neptunian objects are distributed randomly around the Sun, then this will not matter; one patch of sky will be as good as any other. However, if, as seems likely, certain stable regions contain a disproportionate number of objects, the resulting discoveries will be biased depending on whether or not the populated regions happen to lie in the same part of the sky as the galactic plane.

These sorts of observational biases can be estimated and allowed for, but there still remains the problem of actually calculating the sizes of the objects that are being counted. If all that is available is a measurement of how much sunlight is reflected back from an object, its size can only be determined if its reflectivity is known. Without this information it is impossible to tell if one is looking at a small bright object (such as a snowball) or a large dark object (such as a lump of black rock). The reflectivity of planetary surfaces is described in terms of albedo and is 1 (or 100%) for a perfect reflector and 0 for something which is truly and utterly black. Most asteroids have albedoes in the range 0.04 (4%) to 0.16 (16%). While this range might not sound very large it can produce a variation in the estimated diameter of an object by a factor of two and this translates to an error in the mass of a factor of eight. Inactive comet nuclei have albedoes of 2–10% while Pluto, which has a surface covered at least in part by ice, has an albedo approaching 60%. Clearly, before sensible size estimates can be made for the trans-Neptunian objects, astronomers need to decide which albedo value they are going to use for their calculations.

The albedo of an asteroid can be found if simultaneous measurements are made of both the visible light being reflected back from its surface and the amount of heat which it is emitting. This determination is possible because, provided the object is in thermal equilibrium with the incoming sunlight, the sunlight that is not reflected must be absorbed. The energy absorbed goes into warming up the asteroid and is then radiated back into space. The peak wavelength of this thermal emission is determined by the object's temperature; the hotter it is the shorter is the wavelength at which most of the emission occurs. Provided that both the reflected and emitted energy can be measured at the same time, it is possible to use models of the object's likely thermal properties to determine both its size and albedo unambiguously. This technique has been applied to determine the sizes of objects both close to the Earth and in the main asteroid belt. In the inner solar system asteroids have temperatures of between 200 and 300 K and their strongest thermal emission is at wavelengths of about 10 microns. This is a region to which the Earth's atmosphere is fairly transparent and so it can be studied from ground-based telescopes. Objects beyond Neptune are rather colder than this, typically 60 to 80 K. At these low temperatures the bulk of the thermal emission is at longer wavelengths which do not penetrate the Earth's atmosphere and so cannot be measured from the ground.

One stepping stone to the solution of this difficulty is to observe the larger Centaurs, which are closer to the Sun and so are both warmer and brighter. If the Centaurs are objects evolving inwards from the trans-Neptunian region, then their albedoes should be representative of their more distant cousins. The brighter Centaurs have significant thermal emission in the 20 micron region, where there is another atmospheric window for ground-based observations, and thermal infrared observations have been made for Chiron, Pholus and Chariklo. The early observations of Chiron and Pholus were made by a number of people, including Humberto Campins, John Spencer, John Davies and Mark Sykes. They used simple infrared photometers on telescopes such as the IRTF and UKIRT on Mauna Kea where the thin, dry atmosphere is well suited to work in the thermal infrared. Observations of Chariklo were made, also from UKIRT, by Dave Jewitt and Paul Kalas, using a mid-infrared camera called MAX. Jewitt and Kalas obtained just one data point, a measurement of the 20 micron flux, for Chariklo which they quickly published as a letter in the *Astronomical Journal*. When challenged that writing a whole paper around a single data point was a bit excessive, Jewitt was unabashed. 'Ah yes', he said, 'but it's a very good data point'. Jewitt, Kalas and the others agree that combining the infrared data with suitable asteroid thermal models leads to the conclusion that the Centaurs Pholus and Chariklo are very dark, with albedos of around 4%. The albedo of Chiron is rather harder to determine due to the likely presence of coma, which confuses the measurements. It is thought to be a little higher than the other Centaurs, perhaps as much as 10%.

Attempts by a group led by the Max Planck Institute for Aeronomie in Germany to measure the thermal radiation from trans-Neptunian objects using the orbiting European Infrared Space Observatory (ISO) proved very difficult. Although the satellite could observe at wavelengths between 60 and 100 microns, where the objects were expected to be brightest, the ISO telescope had only a small mirror with limited light-collecting power. The telescope also had fairly low spatial resolution at these long wavelengths. Compounding the problems was that the basic long-wavelength photometer turned out to be less sensitive than predicted and so the project had to go ahead with a different instrument mode. This amounted to a camera, comprising a three by three array of very large pixels, observing through a filter centred at a wavelength of 90 microns. The low spatial resolution that resulted meant that it was difficult to determine how much infrared radiation was coming from the object of interest and how much was coming from the background. This further complicated what were already difficult observations.

Unwanted background radiation could come from a number of sources. These included one or more distant galaxies which just happened to fall in the field of view, clouds of interstellar dust within our own galaxy and the warm zodiacal dust inside the solar system. To remove any signal from distant background sources, ISO observed the same region twice, once when the trans-Neptunian object was in it

143

and again when it had moved away. By subtracting the two observations, all that remained ought to have been the flux from the target object. However, this observational technique created an additional complication. The brightness of the zodiacal dust depends on the viewing geometry and, since the Earth had moved between the two observations, the second time around the satellite was not looking through exactly the same column of dust. There was also a possibility, not all that remote as it turned out, that a small main belt asteroid might appear in the beam by chance, further contaminating the data on the distant object. As an insurance against this possibility, each observation was taken in a sequence which included images with ISO's short wavelength camera ISOCAM. The ISOCAM images were taken at a wavelength of 10 microns, where any trans-Neptunian object would be invisible, but a closer and warmer, main belt asteroid would be quite easy to see. It was a wise precaution; one such asteroid, a newly discovered 1 km sized object designated 1997 SU_{15}, did indeed appear to be in the beam during one of the observations.

Despite all these hurdles, ISO does appear to have detected at least two trans-Neptunians, albeit with rather low signal-to-noise ratios. The Plutino 1993 SC, together with its background reference fields, was observed twice. The two sets of observations were made about a year apart. The scattered disc object 1996 TL_{66} and its reference field were observed once. The 1993 SC observations were fairly successful, getting roughly the same answer each time. They suggested that the object had a diameter of about 300 km and an albedo of around 2–3%. Interpreting the 1996 TL_{66} observation was not so straightforward. Although a source with about the expected flux was detected, it appeared in the wrong one of the nine pixels in the camera's field of view. No amount of trying, and the team did try, could explain why the source was not in the central pixel. From these limited observations it was not possible to determine the size and albedo of 1996 TL_{66} with any certainty, but the fact that something which was probably the object was detected at all hinted that it was indeed dark. This conclusion follows from the fact that if 1996 TL_{66} had a high albedo, then its optical brightness would have implied that it was small, and a small 1996 TL_{66} would not have emitted enough infrared radiation to have been detectable by ISO at all!

The confluence of the three lines of evidence from comet nuclei, Centaurs and the ISO measurements, all point to albedoes of around 4% for the trans-Neptunian objects. So, in the absence of anything better, this value is generally assumed to be about right. However, before moving on, let us just sound one note of caution; all the trans-Neptunians may not be the same. As we have seen, the results from optical and infrared photometry suggest that these distant objects have a range of colours and spectroscopic observations show the presences of ices on some, but not all of them. Ices can be very reflective and if 1996 TO_{66} has an icy surface, then its estimated diameter of 750 km might have to be revised downwards, perhaps closer to 300 km. So the adoption of a constant value for the albedo of what may be a very diverse population may be an unwarranted assumption. However, until more definitive data are available, it will have to do for the moment.

In order to tackle the question of the size distribution, various groups have set out to try and measure the cumulative luminosity function of the trans-Neptunian objects. This function, the number of of objects per unit area of sky which are brighter than some specified magnitude, is now fairly well understood for V magnitudes between about 20 and 26. The problems arise at the extreme ends of this distribution when trying to assess the numbers of quite bright, and of very faint objects.

At the bright end there is just one object, Pluto, which shines at a V magnitude of about 14. This is 250 times brighter than the next brightest object in this region of space. Pluto was found by Tombaugh using 1930s' technology and while few people doubt that he did not miss any other bright objects, it is certainly possible that a few fainter ones might have slipped past his tired eyes from time to time. Since there are so many variables in trying to assess just how thorough Tombaugh's search was, and since Pluto may be unusual in a number of other respects (for example it has a satellite and an atmosphere), it is not always included in determinations of the slope of the trans-Neptunian luminosity function. The outer solar system search conducted by Charles Kowal, during which Chiron was discovered, is another survey which ought to be useful in setting limits on the number of bright trans-Neptunians. Kowal covered a fairly large region of sky without finding any very-distant objects, but once again it was a manual search done by blinking photographic plates. Since even slow-moving objects smear out during long exposures, making them that much harder to find, it is also difficult to quantify exactly how sensitive Kowal's search was.

At the other end of the scale, Anita Cochran's attempt to determine the number density of very-faint objects has also resulted in a fair degree of controversy. Cochran's attempts to discover the Kuiper Belt from the McDonald observatory in the early 1990s were foiled by a combination of bad weather and poor seeing. These conditions meant that her programme could not compete with searches being done from better observing sites such as Mauna Kea. Her solution was to go to a telescope where the image quality was guaranteed to be excellent and weather was not going to be a problem. Anita Cochran took the search into space, winning time on NASA's Hubble Space Telescope to search for comet-sized objects beyond Neptune. Her objective was to confirm the link between the Kuiper Belt and short-period comets. She was joined in this effort by Martin Duncan, Hal Levison and Alan Stern.

Thirty four times in August 1994 the Hubble Space Telescope aimed itself at an otherwise undistinguished field in the constellation of Taurus. Especially chosen since it contained few stars and galaxies, the field lay on the ecliptic plane, a prime hunting ground for trans-Neptunians. Unlike most other searches from ground-based telescopes, the field chosen did not lie close to the opposition point. Instead, the telescope was pointed at quadrature, at right angles to the Earth–Sun line. Along this direction the Earth is moving almost directly towards the line of sight and so there is no reflex motion caused as the Earth 'overtakes' distant objects. Instead, any apparent movement of objects seen in the field of view is due to their actual orbital motion. For an object beyond Neptune, this amounts to a little less than 1 arcsecond per hour.

In principle, moving objects could be identified by looking for the tiny streaks they formed as they drifted across the camera's field of view. In practice, since the telescope orbits above the atmosphere, and so is completely unprotected from cosmic rays, images from the HST camera are littered with hundreds of streaks, blobs and hot pixels caused by cosmic rays striking its detectors. The normal technique for removing these defects is to combine many images and rely on the fact that the real objects stay still relative to each other, while the cosmic ray hits appear at random positions across the frame. Unfortunately, while the standard processing techniques would certainly have removed the bad pixels from Cochran's images, they would just as certainly have removed all the objects of interest as well. Instead, Cochran's group adopted a two-stage reduction process. First, they made a normal image of the field which removed the bad pixels and retained all the fixed stars and galaxies. Then they subtracted this final image from all 34 of the original frames, leaving them with a set of frames containing just bad pixels, noise and (hopefully) some distant objects. Next they added together all of the frames using various rates of motion typical for objects in trans-Neptunian orbits. If any such objects were in the field of view, then in the shifted and summed images they would always land on the same pixel and appear as discrete bright spots. The noise and bad pixels would be spread out over numerous pixels and be diluted into invisibility. After making a number of such images, using 154 slightly different rates of motion, Cochran's group examined the results, but found no obvious candidate objects.

Undeterred, they developed an automatic routine which looked for groups of pixels that seemed a bit brighter than the background around them. They then noted these positions as being worthy of further study. Of course, since the objects were only just above the background, it was still possible that they might have been chance alignments of noise. To try and eliminate such false detections, they tried splitting the data up into different sets (all the odd numbered frames, all the even numbered frames, the first 17, the last 17 and so on). They only considered as real those detections which occurred in the same pixel in images made with a number of different combinations of frames. They also tried making images that were shifted in ways that were not consistent with the motion of real Kuiper Belt objects. They searched these images for faint sources in the same way. Since the only thing that was different between the 'real' images and the 'false' ones was the assumed motion of any objects in the fields, these control frames should have provided a measure of the likely number of false alarms caused by random coincidences of noise spikes and other artifacts.

The result of this heroic effort in image processing was that they detected 53 sources in the images processed using realistic rates of motion and only 24 in the ones using demonstrably false rates. Their conclusion was that they had indeed detected a population of small objects at about the distance of Pluto. Assuming the albedo of the objects was the oft quoted 4%, then, with visual magnitudes of around 28, their objects were between 5 and 10 km in diameter. This was exactly what was expected for a population of comet nuclei waiting to be sent towards the Sun by gravitational perturbations. Of course it was not possible to say which of the 53 sources were real comets and which were noise, nor was it possible to use the observations to determine the objects' orbital elements so they could be followed up later. The result was a statistical detection. There were more sources moving in the right sort of way than there were going in the wrong way, so some of them must be real. Or so it seemed.

Doubts about the reliability of Cochran's result soon began to surface since it implied that small objects were tens, if not hundreds, of times more common than would be expected based on the number of larger objects. Hal Levison put up a spirited defence of their result, but two years after the publication of the original paper another appeared with the deceptively banal title of 'An analysis of the statistics of the Hubble Space Telescope Kuiper Belt object search'. In this paper, Michael Brown, Shrinivas Kulkarni and Timothy Liggett considered how many false objects might appear due to various types of random noise and stated quite bluntly that according to their calculations the uncertainties in the number of false objects exceeded by a large factor the number of objects that Cochran and co-workers claimed to have detected. They concluded that 'The detection of comet sized objects in the Hubble Telescope dataset is therefore not possible'.

This was a throwing down of the scientific gauntlet in a quite spectacular manner and it was not long before Cochran's team made a riposte with a paper entitled 'The calibration of the Hubble Space Telescope Kuiper Belt object search: setting the record straight'. In this paper they described how they had implanted artificial objects with a range of brightnesses into their original data and then searched for them using the same methods they had used before. What they found was that the limiting magnitude of the survey, which is generally defined as the level of brightness at which half of the objects known to be in the data are actually found, was a V magnitude of 28.6. As a further check, they brought in another astronomer to perform the same analysis. Peter Tamblyn used software which had been developed independently of the original search team, but his re-analysis produced essentially the same result, a limiting magnitude of 28.4.

Although he has not published a rebuttal of Cochran's rebuttal, Michael Brown remains unconvinced. 'There is no point in continuing to debate this in the literature', he said, 'We want people to look at the papers and make up their own minds as to who is right'. This is good healthy scientific debate and it is how the scientific method works. The key issue is repeatability, not arguments about who knows how to do statistics properly. Based on their earlier success, Levison and colleagues were awarded more time on the Hubble telescope to repeat the experiment with more images. The second set of observations were made in the last week of August 1997 using a somewhat different observing strategy. Instead of staring at a single point and letting the target objects drift across the images, this time they tried tracking the telescope at the typical rate of motion for a Kuiper Belt object. This turned out to be rather more complicated than they expected. As the centre of the field of view moved slowly across the sky, it was necessary for the satellite to switch from one guide star to another as the observations progressed. This made reconstructing the final images very difficult and a lot of time had to be spent just understanding the details of exactly how the Hubble Telescope's pointing system worked. This extra work delayed the processing of the data and by early 2001 no results of the second experiment had been published. Time will tell if the original result will be vindicated.

Whatever the eventual fate of the Hubble Telescope result, the data from the other searches have established a fairly good luminosity function for trans-Neptunian objects with V magnitudes in the range from about 21 to 26. Converting this into a distribution of sizes and masses requires correction for the aforementioned selection effects and a model of how the real objects are actually distributed in space. Most estimates are starting to converge on some values which seem reasonable. There would appear to be about 70 000 objects bigger than 100 km in diameter in the region between 30 and 50 AU, which encompasses both the Plutino and Classical Kuiper Belt populations. That this represents a huge population is illustrated by the fact that there are only a couple of hundred main belt asteroids in this size range. For a reasonable size distribution, the population of 1 km sized trans-Neptunian objects inside about 50 AU is probably about 100 million. Assuming these objects are indeed icy planetesimals, this corresponds to a total mass equal to about 0.2 of the mass of the Earth, a value in encouraging agreement with early estimates, such as those by Whipple, of the likely total mass of the present-day Kuiper Belt.

What of the population of scattered disc objects like 1996 TL_{66}? This is an even trickier problem, since such objects are very difficult to find. By mid 2000, just a handful of such objects were known. A measure of the problem is that one of these objects, 1999 CY_{118}, is bright enough to be detected by a 3.6 m telescope over only 0.24% of its almost 1000 year orbit. Based on these rather small numbers, and some careful simulations of how a population of objects might be discovered by the survey teams involved, Chad Trujillo estimates that there are about 30 000 objects in the scattered disc. This amounts to a further 0.05 Earth mass of material.

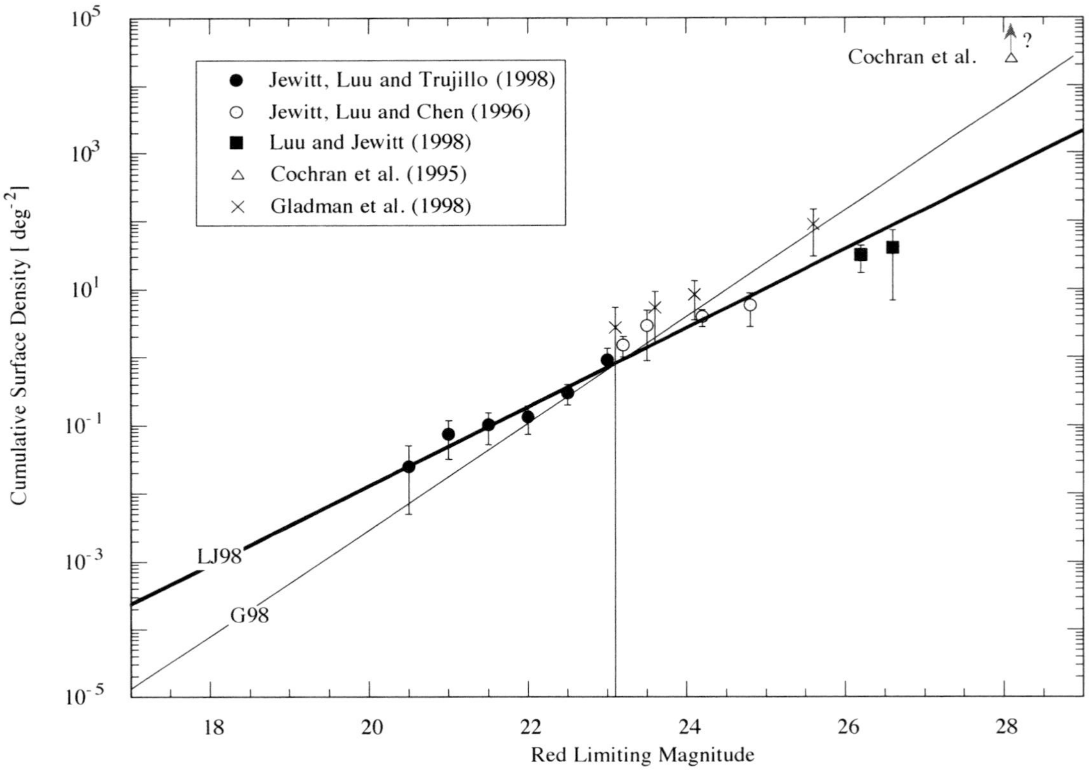

Figure 8.1 A cumulative luminosity plot of trans-Neptunian objects. The horizontal axis is red magnitude with brighter, and so on average larger, objects to the left. The vertical scale is the number of objects found per square degree of sky which are brighter than that magnitude. Data from various groups are plotted and they give roughly the same slope. Note that the point in the top right corner, deduced by Anita Cochran and co-workers using data from the Hubble Space Telescope, is above even the steepest line. (Dave Jewitt.)

Things that go bump in the dark

Analysis of the many surveys of the trans-Neptunian population lead to another interesting conclusion; the Kuiper Belt is quite thick. That is to say that it extends a considerable distance above and below the invariable plane of the solar system. Exactly how thick the belt is is unclear since most of the surveys have been concentrated in the ecliptic region. Objects in orbits with low inclination, in other words those that are orbiting in much the same plane as the rest of the planets, spend all their time near the ecliptic. Even objects in orbits that are inclined relative to this plane still cross it twice on each orbit and spend at least some of their time there. Not surprisingly then, few searches have been done at distances far away from the ecliptic. Thus, the population of objects in highly inclined orbits, which defines the thickness of the disc, is not well known. Despite these uncertainties it is safe to say that the Kuiper Belt is at least 30 degrees thick and that it may actually be much thicker than this.

This thickness has important consequences. Unlike objects all moving around in near-circular orbits in the same plane, like dancers in a ballroom, objects whose orbits cross each other at more drastic angles can collide at quite considerable speeds. Don Davis and Paulo Farinella have modified computer models which they originally developed for studies of the main asteroid belt to investigate the role of collisions in the trans-Neptunian region. Surprisingly, the time between collisions turns out to be roughly the same in both regions; the much larger trans-Neptunian population balances out the larger volume of space involved. The most significant difference is the relative speeds of the collisions. In the asteroid belt a typical collision occurs between two rocky objects at about 5–6 km per second. In the trans-Neptunian region icy objects strike each other at speeds of only 1–1.5 km per second.

Davis and Farinella's computer simulations show that once a population of large trans-Neptunian objects had formed, it was essentially unchanged over the history of the solar system. At the impact speeds involved even a collision between two equal-sized objects is not violent enough to disrupt them if they are more than about 100–150 km in diameter. These larger objects will, however, probably be heavily cratered by numerous smaller impacts. It is also possible that they are not solid bodies, but comprise 'rubble piles' of very large pieces held loosely together by gravity. However, at small sizes the story is quite different. Objects smaller than about 100 km in diameter are broken up by collisions. The collisions produce a cascade of fragments moving at speeds that differ in velocity by tens to hundreds of metres per second compared with the original speed of the parent body. This could allow the fragments to move away from the parent body's original orbit and might well shift them from a gravitationally stable region into one from which they could evolve inwards towards the Sun. If this happened then some of the fragments could eventually become visible as comets. Davis and Farinella estimate that almost all the small trans-Neptunian objects which we see today, and which will eventually enter the inner solar system as comets, are not primordial at all. They are multigeneration fragments of much larger bodies. They also calculate that about ten new comet-sized (i.e. 1–10 km diameter) fragments are being produced every year by collisions in the Plutino and classical Kuiper Belt region. However, the amount of collisional evolution falls off significantly as the distance from the Sun increases. A primordial population of objects at roughly 80 AU (if such a population exists) will not have undergone significant collisional changes since it was formed.

Don Davis admits that astronomers do not yet understand the structure of comet-like bodies well enough to know how they would behave during collisions at velocities of hundreds of metres per second. He and other scientists are working on this problem by carrying out laboratory experiments involving high-speed impacts into icy targets. One of these scientists is Eileen Ryan, from Highlands University in New Mexico. Her speciality is making, and then breaking, snowballs. Ryan says she was fascinated by astronomy from an early age and was always focused on some kind of a career in science. She did an undergraduate degree in physics and, although she started doing postgraduate work in astronomy in New Mexico, it was not long before she changed course slightly. She moved to the University of Arizona, and into planetary science. While a graduate student in Arizona, she found herself working some of the time at the Planetary Science Institute, a small non-profit research group just off the university campus whose staff included Don Davis and several other well-known asteroid researchers. The institute had just started an experimental programme to study asteroid collisions and before long Ryan, who admits to having been at the right place at the right time, was essentially running this project. She found the work interesting and exciting and was very surprised to realise how little was actually known about what happens when two astronomical objects hit each other. Studies of asteroid collisions led naturally to investigating the collisional physics of icy objects in order to understand what was happening in the trans-Neptunian region.

The experimental runs take place at the Ames Research Center at Moffett Field in California. Here, NASA has built a special test facility, the Ames Vertical Gun Range, for experimental work on high-speed impacts. The facility comprises a vacuum chamber, about 2 meters across, in which the targets are placed, and a variety of guns. There are several different kinds of guns which allow projectiles to be fired over a wide range of velocities depending on what the experimenter requests. Targets are installed in the chamber and then bombarded by high-speed projectiles from different angles. Ultrafast cameras, taking

Figure 9.1 Eileen Ryan, whose impact experiments at the NASA vertical gun range are used to help understand how objects in the trans-Neptunian region evolve. (Eileen Ryan.)

400 frames per second and sometimes supplemented by a video camera storing 1000 frames per second, record each test. The Ames gun has been used to study cratering on the Moon and planets by firing projectiles at sand or rock targets, and for studies of collisions between asteroids.

For their tests Ryan and Davis needed targets about 10 cm across, which had to be representative of trans-Neptunian objects. Unfortunately, no-one truly understands what the real things look like or how they are made up. Ryan thinks it is likely that they are porous, fluffy or grainy in structure, but it is not known if all the constituent pieces are of the same size, or if they cover a range of sizes. Accordingly, the experiments were done using a variety of targets which were made using ice blocks bought from a local supermarket. For targets comprising an aggregation of particles with a wide range of sizes, Ryan took some ice chips and hammered them into frag-ments. She then squeezed the fragments together as if making a snow-ball. Other targets were composed only of large pieces, each piece being an ellipsoid about 2.5 cm long (basically ordinary ice cubes). Others were put together with 0.5 cm grains. To make very-small-grained targets, they put ice in a kitchen blender and ground it up before compacting the resulting icy powder. To finish the job, each of the targets was allowed to melt slightly and then pushed into a mould and refrozen. Ryan admits that, 'Making the targets doesn't sound very scientific, but in real life your clever ideas actually have to be implemented somehow'.

A variety of projectile types, including pellets of aluminium, solid ice and fractured ice, are fired into the simulated Kuiper Belt objects. The aluminium pellets are used since similar projectiles had been used in the past for experiments with rock targets and using the same type of projectile simplified comparisons between the different exper-iments. Icy projectiles were more realistic, but rather harder to work with, since the golf-ball-sized ice projectiles tended to break during the launching process. If this happened the projectile would arrive as a stream of ice particles rather than a single object.

Each experimental run lasts about two weeks and comprises thirty or forty shots in total. For each shot Ryan and her group begin by lining the chamber with padding to stop extra unwanted fragmenta-tion caused by pieces flying off the target and then hitting the chamber walls (although experience shows that most fragments do not hit the walls, but just fall to the ground). The chamber floor is covered with a plastic tarpaulin. After the shot, as soon as the air has been pumped back into the chamber, the team rushes in. They scoop up the fragments and then run upstairs to a cold room where the fragments are preserved. They try to collect every fragment weighing more than half a gram, and they usually manage to gather up more than 80% of the original mass of ice. The smaller fragments melt almost immediately and are lost. During the day, or the next day, the fragments in the cold room are measured, weighed and photographed. This allows the fragment sizes and masses, and the mass of the largest fragment to be recorded. The fragment velocity distribution is deduced from examining the film and video recordings taken during each experiment. Ryan takes the results of each shot and uses a two-dimensional hydrodynamic computer code to model the result of the collision and then she compares the outcome calculated by the computer with what actually happens in the test chamber. If the results of the computer model match the data, it is then possible to extrapolate upward by seven, eight or even nine orders of magnitude to estimate the effects of collisions between larger bodies.

According to Ryan, this sort of information is critical to find out how collisions have affected the Kuiper Belt and what is happening to objects in those regions. Astronomers need to know what kind of fragments are produced, what the mass distribution is, at what speeds the

Figure 9.2 A test target representing a planetesimal in the trans-Neptunian region. It was assembled from a bag of commercially available ice cubes. (Eileen Ryan.)

fragments are ejected and what they hit next. The results are interesting, and not very intuitive. It appears that porous ice targets behave as strongly as solid ice during collisions, even though the porous ice targets are weak when subjected to more gradual forces. The reason seems to be that the sudden pulse of energy from the collision is dissipated by the empty spaces within the porous structure. This delays and confuses the shockwave as it travels through the material after the impact. It is a process that is very complicated to model, but it is essential that we understand it if we are ever to understand the collisional evolution of the objects in the outer solar system.

Eileen Ryan's experiments mostly concern collisions in which the target is shattered into many fragments, but smaller collisions must be important too. Given all the other uncertainties, it is not possible to be sure exactly how frequent these collisions are, but it seems likely that a large object, say one about 100 km in diameter, is struck by a 1 km object every 300 years or so. When a smallish object crashes into a much larger one then it is presumably destroyed completely and forms a crater. The impact will excavate material from the target, some of which will land again and form a blanket around the crater. Similar effects are seen in the form of the bright rays which extend outwards from a number of recent craters on the Moon. This impact 'gardening' of the surface (also sometimes called 'space weathering')

Figure 9.3 The fragments of one of Ryan's test targets after an impact experiment. These results suggest that small trans-Neptunian objects may well be irregular. (Eileen Ryan.)

may be responsible for some of the colour diversity reported by various observers making photometric observations. However, since even the largest objects in the trans-Neptunian region are quite small by planetary standards, they do not have particularly strong gravitational fields. Much of the ejected material is probably thrown off into space as dust and fine debris.

This dust will not remain in the trans-Neptunian region forever, or even for very long. A variety of physical effects act on very small particles and cause them to be either blown out of the solar system or to spiral inwards towards the Sun. One of these is called Poynting–Robertson drag. This occurs because sunlight striking a dust grain in solar orbit appears (to the dust grain) to be arriving from slightly in front of the particle. The arrival of this stream of photons on its front face provides a braking force which slows the grain and causes it to spiral into the Sun. Dust is also removed by impacts with other grains already orbiting in the locality or with interstellar particles passing through the solar system. All of these effects conspire to remove dust from the trans-Neptunian region on timescales of around a million years. So while there is probably a cloud of dust in the Kuiper Belt, it is not likely to be very dense.

That any dust in the trans-Neptunian region must be quite thinly spread out is confirmed by the fact that no-one has yet been able to observe it. Fred Whipple rejected searching for sunlight reflected from the dust since it would be too faint to see against the other sources of faint glows in the night sky. More recent attempts have tried to use not light emitted from the Sun and then scattered back towards the Earth, but rather the thermal infrared radiation emitted by the particles themselves. This experiment would be impossible to do from the ground, but in 1990 the American Cosmic Background Explorer (COBE) spacecraft mapped the entire sky in the far infrared to search for cosmological signatures of the Big Bang. Along the way, COBE also mapped out the zodiacal background and numerous other faint sources of infrared radiation. These sources of far infrared radiation had to be modelled so they could be removed from the data before it was searched for the cosmological background which was the main objective of the mission. COBE scientists did not find any evidence for a band of trans-Neptunian dust. While the limit which this failure placed on the amount of dust that might be present was not very strict, it was at least consistent with estimates based on the rate at which dust is being produced by collisions.

To date, the only observational measurements of dust in the Kuiper Belt have come from an unusual source, spacecraft which are speeding away from the Sun after their spectacular encounters with the giant planets in the late 1970s and 1980s. Pioneer 10 flew past Jupiter in 1973 and has since penetrated more than 10 AU into the trans-Neptunian region. The spacecraft's dust-detection system, designed to measure dust in the asteroid belt, is no longer operating, but its hydrazine fuel tank has been used to place a limit on the number of medium-sized particles in the region through which it has been travelling. Scientists have calculated that it would take a particle a few millimetres across to puncture the 42 cm diameter tank and the fact that the tank has survived so far limits the number of such particles. Although the limit which this result provides is not very strict, being about one tenth of an Earth mass, the result is an interesting one since the size range in question is difficult to measure by any other method. Particles in this size range are too large and too far apart to produce detectable diffuse emission and much too small to detect directly. The experimenters have developed their idea to suggest that other parts of the spacecraft, such as portions of the communications system, wiring bundles and some of the other electronics, might also be tested to see if they have suffered impact damage. Unfortunately, the Pioneer spacecraft is now so short of electrical power that there is none to spare. It is no longer possible to turn on selected instruments for even a short time just to see if they still work.

The two larger and more complicated Voyager spacecraft do not have dust experiments, but they have detected trans-Neptunian dust. Both Voyagers are equipped with plasma detectors designed to measure charged particles trapped by the magnetic field of the giant planets. These detectors also register some plasma when a high-speed particle strikes the spacecraft and is vapourised on impact. So, by using the Voyager itself as a detector, and recording the bursts of plasma produced by the impacts, a crude dust detection experiment has been improvised. Of course, it is far from a perfect system. In particular, the range of particles it detects is rather limited; small grains do not produce measurable effects and large ones are very rare. Still, until other spacecraft arrive in this region, it is the best that can be done.

The results of the Voyager experiments are intriguing. One spacecraft is dropping below the ecliptic plane and the other is climbing above it. Both spacecraft did detect dust impacts as they started to traverse the trans-Neptunian region, but in both cases these detections have now stopped. Voyager 2 data cut off at 33 AU and Voyager 1 made its last detection at 51 AU. It may be that the instruments have just ceased working, not unreasonable for equipment that has been in space for close to 25 years, but perhaps the dust has really thinned out to the point that it is not being detected.

Dynamical arguments, sky surveys and spacecraft dust experiments all suggest that the present-day Kuiper Belt contains less than one Earth mass of material. However, there are strong indications that this has not always been the case. The present mass of material may be only a few per cent of the original solar nebula at this distance from the Sun. There are several lines of evidence that this must have been so. The first harks back to the original papers of people like Edgeworth and Kuiper and concerns the overall density of the solar system at increasing distances from the Sun. If it were possible to grind up all of the non-volatile material in the giant planets and spread it out in rings at each planet's heliocentric distance, then there would be a fairly gradual decline in mass density with increasing distance from the Sun until the orbit of Neptune was reached. After Neptune there would be a very sudden and dramatic drop. It was this sharp edge to the solar system that made Kuiper and Edgeworth speculate about a trans-Neptunian disc. Even though such a disc has now been found, the fundamental problem remains. If all the objects and dust in the disc are added together there still doesn't seem to be anything like as much material as there should be.

Another line of evidence comes from the very existence of large Kuiper Belt objects and this has been investigated by the American Alan Stern. Alan Stern is a true aerospace enthusiast. He has degrees in physics and aerospace engineering and is a qualified pilot of both powered aeroplanes and gliders. He graduated with a PhD in planetary astronomy from the University of Colorado and has since been involved in numerous airborne and space astronomy projects. Today he is the director of Space Studies at the South West Research Institute in Boulder, Colorado. Stern has long been interested in the outer solar system. Using computer simulations he has shown that the rate of collisions in a disc with the same mass as the present-day Kuiper Belt is far too slow to grow objects the size of 1992 QB_1. The density of material this far from the Sun is so low that there is simply not enough time since the solar system formed for comet-sized planetesimals to accrete into objects several hundred kilometers in diameter. To reconcile this

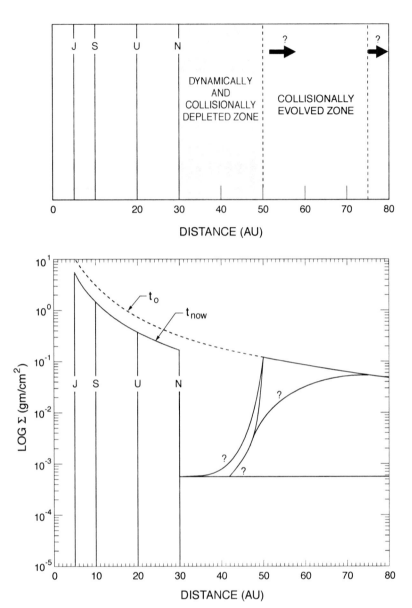

Figure 9.4 The mass density across the outer solar system which would result if the giant planets were ground back up into dust and spread out as rings at their present distance from the Sun. The line labelled t_0 shows the estimated density of the solar nebula early in its history. The solid line t_{now} represents the situation today. Even taking into account the Kuiper Belt objects between 30 and 50 AU, there seems to be a sudden drop in density after the planet Neptune. Alan Stern and others have shown that this region is depleted by erosive collisions, but speculate that beyond 50 AU the density of material may begin to rise again. (Alan Stern/Astronomy Society of the Pacific.)

problem with the fact that such objects clearly do exist, Stern suggests that at some time early in its history the Kuiper Belt must have been more massive. He estimates that the region between 30 and 50 AU must once have contained at least ten, and perhaps as much as fifty Earth masses of material. Interestingly, this is about the amount of material you would get by extrapolating the mass density of the giant planets out into the trans-Neptunian region.

Stern also concluded that even if such a massive disc did exist early on in the solar system's history, the growth of large objects could only have occurred if the objects in the disc were in orbits of fairly low eccentricity and inclination. Only if this was true would the mutual encounters between planetesimals be gentle enough that there would be a good chance that the objects would stick together. The present-day trans-Neptunian disc is quite thick and, as shown by Don Davis and Paulo Farinella, collisions within the disc are more likely to lead to erosion than to growth. So the objects we see there today must have been formed quite early on, before the growth of Neptune and its subsequent gravitational influence on the forming disc could be significant. The existence of numerous Kuiper Belt objects with diameters of a few hundred kilometers suggests a sort of race took place between the planetesimals in the trans-Neptunian region and the proto-Neptune. For perhaps several hundred million years, the planetesimals were trying to grow as quickly as possible before Neptune increased too much in mass. Once Neptune grew to the point that its gravity stirred up the orbits of the nearby planetesimals, their mutual collisions became destructive and any further growth stopped.

Finally, there is the existence of Pluto and its moon Charon. How did this unlikely pair ever form and come into orbit around each other? Pluto and Charon are both small icy worlds. They presumably formed in the outer solar system by the gradual accretion of smaller planetesimals, just like the other objects in the trans-Neptunian region. The problem is that there is no known process which would allow Charon to form in orbit around Pluto. Instead, it appears most likely that two quite large bodies formed independently and then collided. The result of such a collision could have thrown huge amounts of material into orbit around what was left of Pluto and this material eventually re-accreted to form Charon.[†] For a variety of reasons this

---

[†] The Earth's moon is believed to have been formed in a similar way when an object the size of Mars struck the proto-Earth and blasted material into Earth orbit.

is an attractive model but for one thing, the likelihood of such a collision is very, very small. The probability is less than one in a million
over the age of the solar system. Now, million to one chances do come
off, occasionally, but using such rare events to explain things tends to
make astronomers nervous.

One solution to the mysterious origin of the Pluto–Charon binary
is to assume that there were a lot more Pluto-sized planetesimals early
in the history of the solar system. By increasing the number of
objects, the odds of a collision are lowered to something a bit more
reasonable. Alan Stern argued, before a single Kuiper Belt object had
been found, that an ancient population of thousands of 1000 km sized
'ice dwarfs' could be invoked to explain several puzzling features of
the outer solar system. The unusual tilt of the planet Uranus, which
lies more-or-less on its side as it orbits the Sun, could be explained by
an impact by one or more large bodies late in the planet-building
process. Neptune also has quite a large axial tilt, about 29 degrees,
which can be explained in the same way. Neptune has a large satellite,
Triton, which orbits the planet in the wrong direction compared with
the sense of rotation of the planet itself. Triton could be an ice dwarf
that wandered close to Neptune. However, to be captured into orbit
around Neptune, Triton must lose energy. Possible ways this might
have happened include gas drag in the forming planet's outer atmosphere or a collision with a small primordial satellite of the planet.
Either way, Triton has to make a very close approach to Neptune
which is, on the face of it, quite improbable. Every one of the scenar-
ios outlined here are possible in isolation, but the chance that they
would all occur is very low. The fact that they did all occur suggests
that there was once a much larger population of these icy dwarfs. The
existence of this large population greatly increased the probability of
what are otherwise very unlikely events.

What was the fate of the remaining ice dwarfs? Like the many
smaller objects formed in the outer planet region, they were probably
ejected by gravitational interactions with the forming giant planets.
Some may have escaped the solar system entirely, the remainder were
either incorporated into planets or were sent into the Oort Cloud and
the scattered disc. There is no difficulty in doing this; even though the
objects are orders of magnitude larger than typical comet nuclei they
are still ten thousand times less massive than either Neptune or
Uranus and so easy to banish into the outer darkness. The existence of
such objects in the trans-Neptunian region is not ruled out by the

Figure 9.5 Neptune's large satellite Triton. This icy world may be one of the last survivors of a once more numerous population of ice dwarf mini-planets in the outer solar
system. (NASA.)

present data. The cumulative luminosity functions worked out to date
do not rule out the existence of between several and a few dozen Pluto-
sized objects still waiting to be discovered.

All this points to a final question about the Kuiper disc that still
needs to be resolved. What is the density of material beyond about
50 AU, the limit of most ground-based surveys to date? We have seen
that the density of material in the 30–50 AU region is actually a lot
lower than would be expected based on an extrapolation of the density

in the region of the giant planets. This depletion can be explained by ejection of material during the planet-building process and the steady removal of mass ever since as erosive collisions grind down the trans-Neptunian objects. However, beyond about 50 AU the gravitational influence of Neptune is negligible and any planetesimals formed in these distant regions will not have been stirred up by the planets. These distant objects would probably remain in circular, low-inclination orbits where high-speed, erosive collisions are rare and they should still exist today.

Although the mass density beyond 50 AU must have been lower than it was closer to the Sun, if the 30–50 AU region once contained more than ten Earth masses of material, then the outer regions of the protoplanetary disc must also have been quite massive. Simulations of the growth of objects in this region by Stern and his colleague Joshua Colwell have shown that it is possible to grow objects up to several hundred kilometres in diameter in less than the age of the solar system. Although the growth proceeds more slowly so far from the Sun, the lack of any excitation of the orbits by Neptune means that accretion can continue for much longer. Indeed, it may still be going on today. Stern and Colwell argue that since the conditions in the region between 50 and 100 AU from the Sun are not erosive, then the present-day density of material may begin to rise again, quite rapidly, this far from the Sun. If this is indeed the case, then the Kuiper Belt as we know it today may in fact be a local dip in density. There may be a wall of material yet to be found still further out.

To date, no sign of such a wall has been found and increasing numbers of people are beginning to suspect that it never will be. Simulations of the rate at which objects are being discovered show that by now at least a few objects in circular orbits beyond 50 AU should have been found. None have yet turned up. It may be that these more distant objects are on average smaller or darker, and so doubly harder to find, than those closer in. However, according to Chad Trujillo, to explain his deep survey data in this way would require some ridiculous assumptions. The sudden drop in discoveries at 50 AU would require that the more distant objects were all five times darker, or somehow systematically much smaller, than the objects just nearby. R. L. Allen, together with Gary Bernstein, is in broad agreement with Trujillo. Their data suggest that the Classical Kuiper Belt does not extend beyond about 55 AU. In contrast, Brett Gladman feels that since distant objects are so much harder to find he can't rule out a gradual fall off in objects with distance. It may also be that the protoplanetary disc did not extend outwards to great distances after all. Perhaps the disc was truncated by some process. Stars form in clusters and a close encounter between the Sun and one of its siblings might have removed the outer regions of the protoplanetary disc. Luckily, this is not an unanswerable question. Encounters between the Sun and a passing star can be modelled on computers and used to predict details of how the ensemble of objects beyond about 50 AU should appear. In particular, Shigeru Ida, from Japan's Tokyo Institute of Technology, and his collaborators have shown that an encounter with another star would tend to increase both the inclination and eccentricity of the orbits of distant objects. If such distant trans-Neptunian objects exist then once a reasonable number of their orbits have been determined, this hypothesis can be tested.

Dust and discs

The discovery of the Kuiper Belt has established that the planetary region does not stop at Pluto, but that it extends far further into space than originally thought. How does the Sun's trans-Neptunian disc compare with the structures astronomers see around other nearby stars?

Until quite recently astronomers thought that most stars had long since blown away any remnants of the dust cloud from which they were formed and that they now existed in more-or-less splendid isolation. Mature stars lie on what is called the main sequence, a stable period in the life of a star that corresponds roughly to adulthood in humans. However, hints that main sequence stars might not be that simple came during the first few weeks of the mission of the Infrared Astronomical Satellite (IRAS) in 1983. IRAS was an international project to survey the sky in a number of far-infrared wavelength bands that cannot be studied from the ground because of absorption in the Earth's atmosphere. The satellite was launched in January 1983 and, as with any new astronomical instrument, one of the first tasks facing the IRAS team was to check the calibration of the satellite's detectors. The way chosen to do this was to observe a number of stars whose properties were thought to be well understood and whose far-infrared fluxes could be estimated from ground-based observations at other, shorter wavelengths.

One of the stars chosen was Vega, or Alpha Lyra, the brightest star in the constellation of the Lyre. Vega is a well-observed main sequence star which lies quite close to the Sun. It was believed to be of spectral class A0, just an ordinary hot star which ought to behave like a black body with a temperature of about 9850 K. Since the flux from a black body of any given temperature is quite easy to calculate, stars like Vega are often used as calibration sources for astronomical instruments. Vega in particular is used to define the zero point of the astronomical magnitude scale. All astronomical magnitudes in the ultraviolet, visible and infrared regions are linked to the brightness of Vega one way or another. Accordingly, Vega was a prime target early in the IRAS mission. However, quite soon after launch, it appeared that something was wrong; observations of Vega showed that the IRAS detectors seemed to be over-responsive to long wavelengths.

This was a puzzle, and one which demanded the immediate attention of the scientists and engineers at the IRAS ground station at the Rutherford and Appleton Laboratory in Didcot, England. Luckily, it did not take long to find out what was going on. Observations of a number of other sources came out more or less as expected and it was soon clear that the problem was not with IRAS, but with Vega. The star is not a simple black body. It has excess emission in the far infrared which had escaped detection from the ground. Excess infrared radiation is quite commonly associated with young (i.e. pre-main sequence) stars. It is caused by hot dust close to the star which is heated up by starlight and then radiates heat away in the infrared. However, Vega, a well-evolved main sequence star, should have lost all of its original protostellar dust long ago. What was more, the IRAS data indicated that the dust was cool, only about 50 K or so. This low temperature implied that the dust was not close to the star, but that it was in a disc some distance away from Vega itself.

Once the big hint had been dropped by the discovery of the Vega disc, it did not take long before infrared excesses were found around other nearby main sequence stars. Three stars stood out almost at once, Beta Pictoris, Alpha Pisces (Fomalhaut) and Epsilon Eridani. Eventually, a closer examination of IRAS' database showed that numerous otherwise apparently ordinary stars also had infrared excesses, all suggestive of the presence of cool dust. What was more, in almost every case, the age of these stars greatly exceeded the theoretical lifetime of dust in the system. If the dust was just a remnant of the original protostellar cloud from which each star had formed, it should have long since been removed by the same effects that are presently clearing out dust from the Sun's Kuiper Belt. The implication was clear, something must be replenishing the dust around these stars. Could it be mutual collisions within a population of comets and asteroids?

The detectors on the IRAS satellite had very low spatial resolution;

at best it was only a few arcminutes. This was perfectly adequate for a sky survey mission, but was very poor by the standards of a normal ground-based telescope. So, although IRAS could detect the presence of dust around Vega and other stars, it could reveal very little about how the dust was actually distributed. Just about the only conclusion that seemed to be safe was that the temperature of the dust suggested that it was in a disc with a hole in the middle. The next development was not long coming. Brad Smith at the University of Arizona and Rich Terrile at the Jet Propulsion Laboratory in Pasadena, California, began a programme to look at what were already being called 'Vega Excess Stars' using a camera fitted with an optical coronagraph. A coronagraph is a device used to block out the light from a bright astronomical source so that any fainter objects which might be nearby can be seen. The process is very similar to a car driver blocking the headlights of an approaching car with a hand in order to see the much fainter light from the reflectors which mark out the roadway. Coronagraphs get their name from devices used to block the light from the Sun's disc, so that its outer region, the corona, can be studied, but they work just as well during regular astronomical observations looking for faint companions to bright stars.

Smith and Terrile turned a coronagraphic camera towards the star Beta Pictoris (usually abbreviated to Beta Pic) and hit the jackpot almost at once. Clearly visible on either side of the star were wings of reflected light. Beta Pic had a disc of dust and gas almost exactly edge on to us. Particles in the disc were scattering the visible light from the star back towards the Earth. The disc was huge, spanning 1500 AU from edge to edge. Closer examination of the images showed that the disc was not symmetrical and that it was quite thick. The thickness hinted that there might be planetary-sized bodies embedded in the disc which were stirring up the material, stopping the disc flattening out.

Since Smith and Terrile's discovery many more observations have been made of the Beta Pic disc. During 1993 Paul Kalas and Dave Jewitt investigated the asymmetries in the disc in detail and showed that the dust orbiting between 150 and 800 AU from the star was indeed asymmetrical. They concluded that the large scale of this asymmetry meant that it was unlikely to have been caused by planets orbiting close to the star. They suggested instead that the disc may have been disturbed by the gravitational field of a star passing close to Beta Pic sometime in the last few thousand years. More recent observations have verified the large-scale asymmetry of the disc confirming that it is brighter on one side than on the other. Observations from the Hubble Space Telescope also revealed the presence of warps or kinks in the inner regions of the disc. These inner structures are on scales similar to the dimensions of our own solar system. They point even more clearly to the existence of planetary bodies orbiting somewhere inside the system and gravitationally modifying the structure of the disc.

Optical images of the other Vega excess stars did not detect any more discs, but this was not entirely surprising. For a disc to be seen in normal reflected light, it must lie almost edge on to the line of sight. Such a situation is likely to be quite rare. Circular discs seen close to face on are very difficult to detect directly, since there is not enough material along each sightline to reflect much light back towards the Earth. The best way to detect nearly face on discs is not to rely on reflected starlight, but rather to observe at wavelengths where the warm dust is actually emitting radiation. The dust disc of a typical Vega excess star has a temperature of about 40–70 K. The peak of the thermal emission from such dust is in the far infrared, but there is

Figure 10.1 Space telescope images of the Beta Pictoris Disc at two different scales. The top image shows the large-scale structure of the outer disc, the bottom one shows warps in the very central regions which may result from the gravitational effects of planet-like bodies close to the star. (NASA/STScI.)

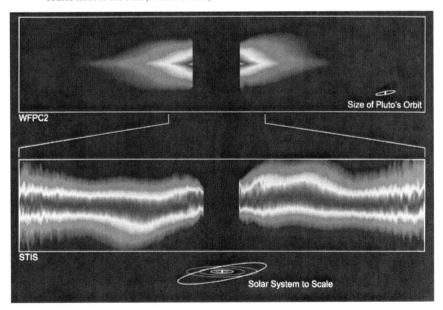

still significant flux in the somewhat longer sub-millimetre region. By good fortune, this area of astronomy was just being opened up to ground-based astronomers.

Millimetre and sub-millimetre telescopes are large dishes rather like radio telescopes in appearance. The main difference is that since the sub-millimetre band encompasses much shorter wavelengths than the traditional radio bands, sub-millimetre telescopes have surfaces that are much smoother than their radio cousins. Sub-millimetre radiation is strongly absorbed by water vapour and so astronomers interested in this wavelength region have sought high and dry locations for their instruments. One such sub-millimetre telescope is the James Clerk Maxwell Telescope (JCMT) on Mauna Kea. The JCMT has a dish 15 metres in diameter comprising of 276 reflective panels. All of the panels are aligned and supported so that the dish is very smooth; deviations across the surface are no greater than the thickness of a page from a typical telephone directory. The whole telescope is inside a weather-tight enclosure which protects it from the elements and maintains the critically important shape of the dish.

Sub-millimetre instruments, usually called bolometers although their function is to do photometry, were soon trained on the Vega excess stars in an attempt to map out the suspected dust shells at long wavelengths.[†] Observations were made from the JCMT and from other similar telescopes such as the Institut de Radioastronomie Millimétrique (IRAM) in France and the Swedish-ESO Sub-millimetre Telescope (SEST) in Chile. Unfortunately, the first results were not very encouraging. Sub-millimetre observing was in its infancy and like early optical and infrared photometers, each instrument had just a single detector viewing the sky. This made mapping extended emission a very laborious process. To make a map it was necessary to first point the telescope at the star, then move off slightly to one side before making a measurement. Once this was done the telescope was pointed to a position on the other side of the star and the process was repeated. Given enough time, and enough good weather, a small grid of positions could be mapped around the star. A crude image could then be reconstructed by computer processing of all the data. It was a slow and tedious procedure. Mapping faint sources was limited by both the relatively low sensitivity of the early instruments

---

[†] Most textbooks define a bolometer as a device which absorbs radiation of all wavelengths. By placing filters in front of it it can be used as a photometer.

and by the large beams, and so poor spatial resolution, produced by even the best sub-millimetre telescopes. Worse still, several groups trying to make similar observations did not get similar results and there was some finger pointing as each group tried to defend its own measurements at the expense of those from its rivals.

This situation was only resolved (if the reader will pardon the expression) by the development of new sub-millimetre instrumentation with multiple detector elements of improved sensitivity. Probably the best known of these instruments is SCUBA, the Sub-millimetre Common User Bolometer Array, installed at the JCMT in Hawaii. SCUBA has over 100 bolometer detectors, each of which is about ten times more sensitive than those in the previous generation of instruments. Since the signal-to-noise ratio of an observation increases as the square of the sensitivity, and in direct proportion to the number of detectors, SCUBA is about 10 000 times more efficient as a mapping instrument than any of the single-element bolometers which preceded it. SCUBA was an obvious tool with which to attack the problem of the Vega excess stars and it was not long before several groups attempted to do just that.

The first and most spectacular success fell to the wife and husband team of Jane Greaves and Wayne Holland, staff members of the Joint Astronomy Centre in Hawaii. They used SCUBA to map the sub-millimetre emission from a number of the brightest Vega excess stars and soon came up with a very interesting conclusion. Their images showed that not only did these stars have the dusty discs predicted by the far-infrared observations from IRAS, but the discs were asymmetric. They had blobs of emission around the circumferences as well as clearly defined holes in the middle. SCUBA was a new instrument and one which had been quite difficult to bring into regular use. Since the presence of asymmetries in the discs was unexpected, considerable effort was put into checking that the features in the SCUBA images were real and not an artifact of either the new instrument or its complicated data reduction software. After careful checking, Greaves and collaborators were convinced; the structures were real.

The SCUBA images immediately explained why the first sub-millimetre observations had been so difficult to reconcile. Different telescopes, with their inevitably different beam sizes, had seen different parts of these structured discs. Each had detected different amounts of radiation. As Jane Greaves put it, 'If only they had looked further out they would have seen lots of flux from the outer regions of the discs, but of course they didn't know that there was anything out there. They were basically staring through holes in the central regions of the discs and missing most of the flux'. The SCUBA images, coming as they did about the same time that great strides were being made in understanding the outer regions of the solar system, suddenly made perfect sense. SCUBA, and the other sub-millimetre cameras coming into service, were staring at systems which had apparently already started to form planets. Each star was surrounded by its own population of dust and planetesimals. Even the scales of the discs were about the same as the Sun's Kuiper Belt.

The most spectacular extrasolar dust disc is that around the nearby star Epsilon Eridani. This star is about 0.8 times the mass of the Sun and is estimated to be about as old as our own Sun was during the final stages of planet formation. Epsilon Eridani seems to have a dust ring extending between about 35 and 75 AU and containing at least 1% of an Earth mass of material. This is actually a rather small amount of dust, compared with the Sun's Kuiper Belt, especially for a disc which is comparatively new and has not yet lost much of its material. Jane Greaves says this is not an insuperable problem. 'SCUBA is sensitive to quite small grains', she says. 'SCUBA sees mostly particles about 100 microns, or one tenth of a millimetre, in diameter. If the disc around Epsilon Eridani has evolved to the point that it contains significant numbers of centimetre or large sized particles, there could easily be a lot more mass than we think. The problem with larger particles, at least as far as SCUBA is concerned, is that they have a lot of mass, but only a relatively small surface area. Large particles like these do not radiate all that much energy in the sub-millimetre and this makes it hard for SCUBA to see them.'

What of the structure seen in the Epsilon Eridani dust disc? The cavity can be explained by the presence of planets whose gravity has swept up the residual dust and cleared out the inner disc. Our own solar system is almost devoid of dust inside the orbit of Neptune for much the same reason. The outer region of the Epsilon Eridani ring comprises a structure like the Sun's Kuiper Belt where mutual collisions between small planetesimals are replenishing the tiny dust grains and making the disc detectable with SCUBA. As for the asymmetries in the disc, these may be huge clumps of material which are being concentrated by one or more forming planets whose gravity is still sweeping up the orbiting dust. Further evidence that Epsilon Eridani may have a planetary system was announced in the middle

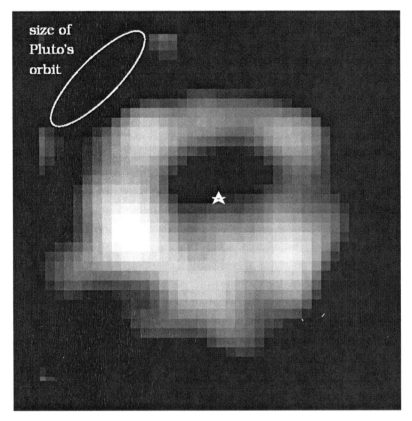

size of
Pluto's
orbit

Figure 10.2 A SCUBA image of the dust ring around Epsilon Eridani. The ring appears to have a hole in the centre and clumps of material around its circumference. The extent of the ring is similar to the Sun's Kuiper Belt. (Jane Greaves.)

of 2000. While Anita Cochran had been searching for small trans-Neptunian objects around our own Sun, her husband Bill had continued with a long-running project to search for large extra-solar planets. Cochran, and other competing groups studying other stars, did this by looking for tiny back-and-fro motions caused by the gravitational tugs from unseen planets around nearby stars. After some twenty years of monitoring Epsilon Eridani, Cochran believes there is strong evidence that at least one large planet orbits the star. The planet is about half the mass of Jupiter and has an eccentric orbit which takes it around the star every 6.9 years. Cochran's discovery adds to the belief that Epsilon Eridani today may be a snapshot of what our own solar system looked like a few billion years ago.

Although Epsilon Eridani is the most spectacular of the dust rings so far detected by SCUBA, it is not the only one. Observations of Vega have also detected the disc around this star. The Vega disc is asymmetric, implying the presence of planets, or at least the progenitors of planets, around this star as well. The SCUBA images also solved one of the mysteries surrounding the early sub-millimetre observations of Vega. Maps made with single element bolometers had apparently correctly detected that one side of the suspected disc was brighter than the other, but the data were so uncertain that it was hard to be sure. Images from SCUBA confirmed the presence of significant disc asymmetries which were consistent with the much less reliable, but ultimately correct, measurements made almost a decade earlier.

About the same time as the publication of the SCUBA images, two American teams announced direct evidence for a dust disc around another star. Both of these groups were using cameras operating in the mid-infrared region, that is to say wavelengths of around 10 and 20 microns. As with the sub-millimetre band, it is only recently that mid-infrared instruments have progressed from single-element detectors to cameras containing detector arrays with many pixels. In March 1998, two such instruments were turned on the star HR4796A. These cameras, called OSCIR and MIRLIN, were fitted with a 128 pixel square detector array developed by the Boeing company. MIRLIN was mounted on one of the Keck telescopes atop Mauna Kea and OSCIR was fitted to the 4 m Blanco telescope at the Cerro Tololo Inter-American Observatory in Chile.

HR4796A is an A0 star in the southern constellation of Centaurus. With a V magnitude of 5.78 it is just visible to the naked eye. The star is believed to be about 10 million years old, quite young compared with the Sun, but about the right age to be in the late stages of planet formation. It lies 67 parsecs (about 210 light years) from the Sun. This is about three and half times further away than Beta Pic and the extra distance makes detailed observations of it that much more difficult. HR4796A is part of a binary star system and has a dwarf companion about 500 AU away from it. HR4796A was known to have a significant infrared excess, and so was a good candidate to have a circumstellar disc. As in the case of Vega, the temperature inferred from IRAS observations suggested that the disc did not extend all the way to the star, but was confined to a ring between about 40 and 200 AU.

The two observing teams reported similar results, both of which appeared in the same issue of the *Astrophysical Journal*. The observations show that HD4796A has an almost edge on disc which extends out to about 110 AU from the star. The inner edge of the disc cannot be

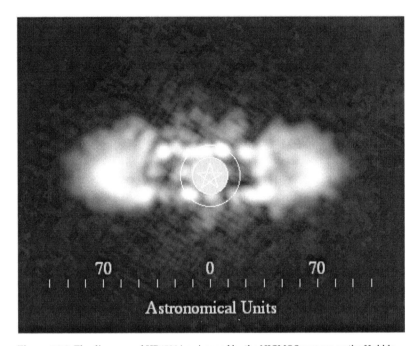

Figure 10.3 The disc around HR4796A as imaged by the NICMOS camera on the Hubble Space Telescope. The central region was blocked out by a coronagraph to suppress the light from the star, revealing the narrow disc. Sophisticated image processing of light escaping around the edge of the coronagraphic disc was used to reveal structure close to the centre of the image, resulting in a slightly different appearance of these regions. In reality, the disc is probably the same all the way around. (Glenn Schnieder.)

detected on the images directly, but since the dust appears to be at a temperature of about 110 K, the observers estimated that the inner boundary was located about 50–55 AU from the star. These mid-infrared observations were soon complemented by observations from the near-infrared camera NICMOS on the Hubble Space Telescope. NICMOS had a coronagraphic mode, which made it suitable for observing faint structures near to bright sources, and its operation at shorter wavelengths and location above the atmosphere offered much higher spatial resolution than that available from ground-based mid-infrared cameras. The NICMOS images showed that the ring was centred about 70 AU from the star and that it was quite narrow, only about 14 AU across. The images also confirmed that there was little material inside about 45 AU.

The outer edge of the dust ring around HR4796A is probably constrained by the gravitational effects of the star's companion. Although small, the companion star's gravity is quite capable of stripping away dust which moves too far from HR4796A. However, the inner edge of the disc cannot be explained in this way. It must be controlled by something interior to the disc. Although the hole might be cleared by effects such as Poynting–Robertson drag, which would cause the material to spiral in towards the central star, an obvious explanation is the presence of one or more planets orbiting inside the disc. This 'Planets plus Kuiper Belt' model is supported by three other factors. Firstly, both infrared and sub-millimetre observations suggest that the mass of dust around HR4796A is somewhere in the region of one Earth mass, about right for a Kuiper Belt. Secondly, JCMT observations suggest that there is not much molecular hydrogen gas in the system, which means that it is no longer possible to form a Jupiter-sized gas giant planet in the system. If HR4796A is fated to have such planets, they must already have formed. Thirdly, the MIRLIN observations point to additional material at temperatures of a few hundred kelvin, typical of a source within a few astronomical units of the star. This is just about the location of the bulk of the zodiacal dust in our own solar system. Taking all these observations together, HR4796A looks rather like a system mid-way between the massive discs around young pre-main sequence stars and the more mature systems like Beta Pic and Epsilon Eridani.

What may well be another extrasolar Kuiper Belt was imaged from NASA's IRTF telescope by David Trilling and Robert Brown. They used a cooled coronagraph (CoCo) together with the IRTF's near-infrared camera to observe 55 Cancri, a Sun-like star believed to be about 3 billion years old. The existence of a planet a bit larger than Jupiter orbiting 55 Cancri had already been deduced from measurements of tiny wobbles in the star's motion and Trilling and Brown set out to see what else might be there. By blocking light from the star itself, the coronagraph allowed them to image a region extending as close as 18 AU to 55 Cancri. Their images revealed an ellipse of scattered light coming from a disc of material extending out at least 40 AU from the star. The observations were made through three different filters in the 1–2.5 micron region and the different appearance of the disc in the different filters gave a clue to the composition of the material present. In particular, while fairly bright at wavelengths of 1.62 and 2.12 microns the disc was almost invisible at 2.28 microns. Since methane absorbs light of this latter wavelength, this suggests quite strongly that the material in the disc has methane ice on its surface.

Trilling and Brown imaged a number of other stars to check that the extended light around 55 Cancri did not arise from some instrumental defect or error in their observing plan. The other stars did not show any similar features, giving considerable confidence that the excess light around 55 Cancri was indeed due to a Kuiper Belt like disc of material. Assuming that the disc was roughly circular and flat they were able to deduce, from the shape of the ellipse in their images, that the disc was tipped about 27 degrees to the plane of the sky. In other words they were seeing it close to, but not quite, face on. They concluded that 55 Cancri had a mature solar system containing at least one large planet and a primordial disc.

The SCUBA, MIRLIN, OSCIR and CoCo observations bring us full circle. They seem to show that dusty and icy discs are common around main sequence stars and that planet, or at least planetesimal, formation is a common process throughout the galaxy. The discovery of a number of other extrasolar planetary systems in recent years has added to the belief that the solar system is no longer a unique flash in the pan. It seems that planet formation is a common process which occurs around many normal stars. It is an interesting confluence of many apparently unrelated lines of enquiry that has brought together stellar astronomers and planetary scientists to provide this huge leap in our understanding of both our own solar system and of planetary systems around other nearby stars.

Where do we go from here?

CHAPTER 11

Since it is a subject which is developing so quickly, it is difficult to predict the future of Kuiper Belt research in detail. None the less some general directions are clear. Astronomers need better mathematical models of how the Kuiper Belt was formed and evolved, and they need to understand the chemical composition and physical characteristics of many more individual objects.

To theoretician Martin Duncan the future direction of his work is clear. He says that physicists have to understand in detail how the process of planet formation is intertwined with the final orbits of the asteroids and comets that we see today. This will require a clearer picture of how planets accrete from myriads of kilometre-sized planetesimals in the solar nebula. Progress will depend on new and ever more detailed computer models able to account for, and track, the mutual interactions of huge numbers of test particles. These new models will have to include the effects of gas dynamics on the evolution of the nebula, extra detail which will add greatly to their complexity. Duncan and his ilk will also need increasingly sophisticated computer algorithms to model the collision and fragmentation of their test particles. Developing these models will bring together the very abstract studies of the celestial mechanicians with the practical work of the snowball smashing experimentalists like Eileen Ryan.

Duncan envisages that five to ten years from now he and his co-workers will be using parallel clusters of computers each many times faster than today's machines. The computers will spend their time doing fairly detailed simulations of the formation of planets and their migration across the still-evolving solar system. The results will reveal how gravitational scattering led to the formation of the Oort Cloud and will illustrate the gravitational sculpting of the Kuiper

Belt. The validity of the models will then be tested by comparison with the statistics of real comets and Kuiper Belt objects until most, if not all, of the uncertainties in the models have been removed.

One project which will help to establish some of those observational constraints is being set up in the mountains of central Taiwan. The Taiwan–American Occultation Survey (TAOS) is an international collaboration conducting a census of the comet-sized objects in the Kuiper Belt. Such small objects are too faint to be detected directly, so TAOS will count them another way. Night after night its telescopes will monitor a few thousand bright stars and watch for any sudden dimmings caused as Kuiper Belt objects pass in front of them. Each occultation will be brief. It will take less than a second for the shadow of a Kuiper Belt object a few kilometres in diameter to sweep across one of the TAOS telescope sites. To be sure that the reported dimmings are real events and not noise there will be several telescopes set along a line. Real occultation events will be seen in turn by each of the telescopes situated along the path of the object's shadow.

Since the detectability of occultations is determined by the brightness of the background star, not by the brightness of the Kuiper Belt object, the TAOS survey does not require large telescopes. Instead, it needs to be able to monitor as many stars as possible as frequently as

Figure 11.1 A schematic diagram showing how the Taiwan–American Occultation Survey will make a census of small trans-Neptunian objects. As an object crosses in front of a distant star it will cast a tiny shadow onto the Earth. If the shadow passes across the TAOS telescopes, it will be recorded. (Lawrence Livermore National Laboratories.)

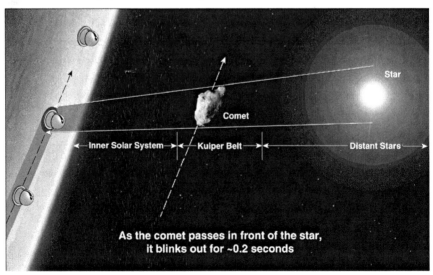

possible in order to maximise the chances of seeing the relatively rare occultation events. The project will start with just three telescopes, each 0.5 m in diameter, situated in the the Yu-Shan (Jade Mountain) National Park in Taiwan. Each telescope will be equipped with a 2048 by 2048 pixel CCD camera covering an area of about 1.7 degrees of the sky. The telescopes will be separated by 5–10 km along a line that runs roughly east to west. They must be located sufficiently far apart that chance atmospheric fluctuations, clouds and so on, do not affect the data from more than one telescope, but the spacing cannot be too large. If the separation is too great then the shadows of the Kuiper Belt objects, which are unlikely to be moving exactly along the east–west direction, will fail to pass over all of the telescopes. If this happens the events will not be recognised.

The telescope sites are very isolated, so each observatory must operate automatically. Power will be generated by solar panels during the day and then stored in batteries until it is needed. Each telescope will be enclosed by a simple box-like structure topped by a folding clamshell cover. The cover will fold down out of the way when the tele-scope is ready to start observing. To protect the telescope against possible power failures, the cover will be counterweighted and will close automatically if needed. Once the system is running, the TAOS project expects to monitor 3000 stars and to produce about 100 000 million photometric measurements a year. Out of these anywhere from a few dozen to a few thousand occultations must be detected, confirmed by cross checking with the other telescopes in the network and communicated to the handful of people who are actually running the experiment. It will be up to them to understand what the results mean in terms of the population of the trans-Neptunian region.

However, in parallel to modelling how the solar system might be evolving and studying the statistics of faint Kuiper Belt objects, astronomers need to be cataloguing more objects to constrain these very models. Dave Jewitt is very clear about this and speaks elo-quently for the need to find many, many more members of the trans-Neptunian population. One of his favourite themes is to draw an analogy between the Kuiper Belt and the rather better understood asteroid belt between Mars and Jupiter. The first asteroid was found in 1801 and, within a couple of years, three more were discovered. At the time these were discovered, almost nothing was known about their physical properties and no-one really understood what they were doing there. The rate of asteroid discovery remained slow for almost a century until new technology, notably the use of photographic plates, was applied to the search. Once photography was introduced, many more objects began to be discovered and patterns in their orbits began to emerge. Gradually, these discoveries led to the realisation that orbital resonances (mostly connected with nearby and massive Jupiter) were important in determining the structure of the present-day asteroid belt. As still more orbits were determined it was realised that there were clusters of asteroids with very similar orbits. These clusters are attributed to families of objects formed by the break-up of larger asteroids, whose fragments are still travelling around the Sun in much the same orbit as their parent bodies.

Even so, it was not until the 1970s and 1980s that the full complexity of the main asteroid belt was understood. Only then did a clear picture of the relationship between asteroids in the main belt and planet-crossing objects and meteorites emerge. The key to this was better computers and, most of all, the determination of accurate orbits for ever more objects. It was these large databases which made possible detailed statistical studies of the evolution of the asteroid population as a whole. Together with better ground-based observations, and recently, images from spacecraft flying past and even orbiting asteroids, the whole complex picture of the history of the asteroid population was slowly put together.

The observational situation with the Kuiper Belt is very similar to that of the main belt a hundred or more years ago. Astronomers have detected only a tiny fraction of the objects that must be out there, and they have determined reliable orbits for only a few dozen of those. Just as two centuries ago, new objects are being discovered and then being lost again due to lack of astrometric follow up. The discoveries so far suggest that the trans-Neptunian region is far more complex than anyone ever guessed, and that its structure probably holds important clues as to how the solar system formed. So, if we are ever going to understand this region of space, we need reliable orbits for thousands of objects. This information will make possible detailed analyses of the populations of the different resonances and of the size distributions of the objects in each region of the Kuiper Belt. To Jewitt, the choice is very clear. We can either fool around for another hundred years or so discovering a few dozen objects a year, or we can do the job properly, find and catalogue ten thousand objects in a single project and solve all of the dynamical problems in one go.

Dave Jewitt believes that what is needed is some kind of dedicated

Kuiper Belt telescope, or at least a sky survey telescope which has discovery and follow up of trans-Neptunian objects fair and square amongst its objectives. At a workshop held at the Lowell observatory in Flagstaff during the autumn of 1998, Jewitt described one such instrument. He envisaged a 4 m telescope with a field of view at least a degree in diameter feeding a mammoth CCD array. Each dark night the telescope would scan the sky for moving objects, generating perhaps 100 gigabytes of data each night. Manual searches of this huge dataset would be impossible, so moving-object software, based on that already being developed and used by people like Chad Trujillo and those at Spacewatch, would search the data. The software would identify likely candidates and pass that information back into the search programme. This real-time feedback would ensure that subsequent observations would be adequate to determine reliable orbits for each discovery. Nothing would be lost. After a while, the necessary follow-up observations of hundreds of objects would take a significant fraction of the total observing time. Hopefully, within a decade or so, ten thousand new trans-Neptunian objects would have been found and tracked in detail. If successful, this effort would reveal the dynamical structure of the outer solar system once and for all.

It is not clear if such a Kuiper Belt telescope will ever be built, but a number of studies of large survey telescopes are now underway. One of these concepts is for a large telescope dedicated to solar system astronomy to be installed on Mauna Kea. The new telescope would replace NASA's aging 3.5 m Infrared Telescope Facility. Like the IRTF, this New Planetary Telescope, or NPT, would be designed to support a range of NASA projects and would spend most, or all, of its time on planetary observations. The proposed NPT would be used for studies of objects being targeted by NASA space missions and to search for previously unknown asteroids and comets. In particular it would look for objects which pass close to the Earth and would make a survey of the trans-Neptunian region. For surveys, the telescope would have to operate in a wide field mode, perhaps covering about 4 square degrees in a single pointing. To detect 10 000 trans-Neptunian objects brighter than about 24th magnitude would require a search of about 5000 square degrees of sky, something which might be accomplished in a year or so with the proposed telescope. Astrometric follow-up of these objects to determine their orbits would require additional time over the next few years. The time needed for follow-up is why a dedicated planetary telescope is needed. With so many other interesting astronomical projects to do, long blocks of time on large telescopes are generally hard to come by and yet without them much of the effort spent on searches is wasted.

However, a suitable planetary telescope would not be limited to just surveying for new trans-Neptunian objects. To accomplish other aspects of its mission the telescope would have to be designed to have a very-high-resolution imaging mode, albeit over a much smaller field of view. This would mean that it could also be used for studies of individual objects. Potential projects might include searches for trans-Neptunian objects that, like Pluto, are binaries. This would require a telescope able to go very deep, in order to detect any faint companions, and to spend considerable amounts of time monitoring the binary candidates to determine their orbits. However, if binary systems in the Kuiper Belt can be detected, then the potential benefits are enormous. Once the orbital period of the system has been found, then the masses of the components can be determined quite easily. This in turn allows the density of the objects to be estimated. The densities provide important information on the physical make-up of the objects. For example, it may reveal if the objects are solid or just loose aggregates of icy boulders. This structural information can then be related to the outcomes of laboratory impact experiments and mathematical models of the fragmentation process.

At least one other exciting prospect is on the horizon for astronomers interested in the Kuiper Belt. NASA plans to send a space mission to Pluto which, if all goes well, will go on to encounter one or more trans-Neptunian objects. Like most space projects, the mission has already been through a number of evolutionary changes as NASA's long-range plans have been modified by budgetary and political factors. Although it is still part of the planned NASA programme, it is by no means impossible that circumstances will force further changes to the detailed mission profile.

NASA first considered sending a spacecraft to Pluto as part of a so-called 'Grand Tour' of the outer solar system. The Grand Tour would have involved a number of large and complex spacecraft launched during the 1970s. These ambitious, and very expensive plans were soon scaled back. They evolved into the highly successful Voyager project which sent two spacecraft to Jupiter and Saturn. As part of an extended mission, Voyager 2 went on to explore Uranus and Neptune, but the flight paths of the two spacecraft were such that neither of them could be diverted towards Pluto without compromising another

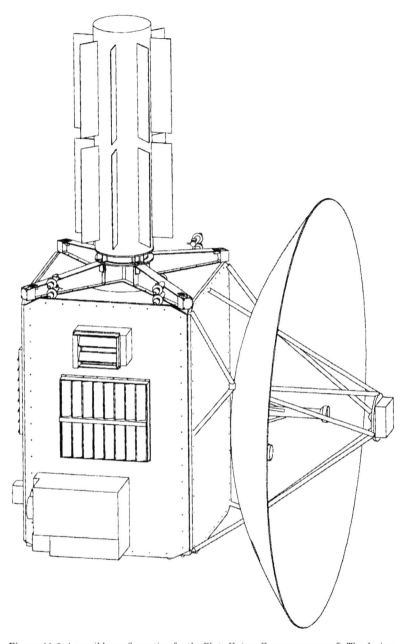

Figure 11.2 A possible configuration for the Pluto Kuiper Express spacecraft. The design is simple and functional to reduce costs and minimise the likelihood of technical failures. (Rich Terrile, JPL.)

important scientific objective. Accordingly the outermost planet remained unvisited. The idea of a mission to distant Pluto then fell from favour for a while, before resurfacing in the late 1980s.

By 1989 NASA was considering a number of possible new solar system missions, including ones to explore Neptune and Pluto. These missions included projects based around the so-called Mariner Mk II spacecraft, a large multi-purpose craft capable of being used for a variety of different deep space missions. However, support for space exploration was dwindling and the NASA budget was shrinking steadily. The high cost of large planetary missions made the prospects for a Pluto mission, especially one likely to take twenty years to complete, look grim. So, in 1991, a new concept began to take shape. The idea originated during a ceremony at NASA's Jet Propulsion Laboratory to mark the issue of a set of stamps commemorating America's planetary exploration projects. All the planets, and the Earth's Moon, were represented with a picture of both the planet and one of the spacecraft which had visited it. The exception was Pluto, whose stamp showed just an impression of the planet and the words, 'Not yet explored'.

The Pluto stamp crystallised the frustration of some of the people working on proposals for a Mariner Mk II class Pluto mission. Deep down they knew that a large and complex mission to Pluto would never be approved and so they began to consider other options. Soon the idea of a small, low-cost Pluto mission dubbed alternately, 'Pluto Very Small' or 'Pluto Fast Flyby' started to emerge. The idea was first presented to NASA in 1992 and it eventually reached the office of Daniel Goldin, the newly appointed NASA administrator. The proposal was timely, Dan Goldin was trying to move NASA towards a new 'faster–cheaper–better' philosophy. He believed that the future of planetary exploration lay in NASA developing numerous small projects, not in building a few large and expensive spacecraft every decade. To keep costs down, each of Goldin's new generation of small missions would be aimed at investigating a limited number of scientific questions on a rapid timescale.

In 1994 NASA decided that even the low-cost Pluto Fast Flyby was still too expensive and it was decided to try again with an even cheaper mission. This concept became known as the Pluto Express and was intended to send two small spacecraft to Pluto. Each spacecraft would have been launched separately and they would begin their journey by heading towards the giant planet Jupiter. Swinging close to Jupiter, the spacecraft would use the giant planet's gravity to pick up speed and be flung towards Pluto. They would arrive there, a few months apart, about 10 years after launch. A scientific definition of the mission prepared in 1995 described a series of studies to be carried out at Pluto and mentioned a possible mission extension to fly past a Kuiper Belt object. Given the large number of potential targets in the trans-Neptunian region, these additional flybys did not present too many technical difficulties. Unless there were severe problems *en route* to Pluto, both spacecraft were expected to have sufficient fuel left to manoeuvre themselves towards suitable Kuiper Belt objects once their primary mission was complete. Inevitably, the mission continued to evolve and an early casualty was one of the two spacecraft. A two-spacecraft mission offered the chance to do additional scientific observations of the Pluto–Charon system and, of course, provided a safety margin in the event that one of the two craft failed completely. However, money was tight and NASA could not afford to do everything it wanted. Faced with a choice between sending two missions to Pluto, or one mission to each of two different destinations, the decision was made to restrict the Pluto mission to a single spacecraft and to save money that could then be used elsewhere.

Although the possibility of an extended mission had been in the minds of planners for some time, the increasing interest in the Kuiper Belt meant that eventually the mission was renamed. The mission is now known as the Pluto–Kuiper Express, although its primary aim remains the exploration of the Pluto–Charon system. Key scientific objectives are characterising the surface geology of Pluto and Charon, mapping the surface composition of both objects and studying Pluto's tenuous atmosphere. Although the details of the spacecraft and its trajectory are likely to change between now and the launch date, it is already possible to sketch out roughly what the mission profile will look like.

One scenario developed in the late 1990s envisaged a launch on a conventional rocket in December 2004, a flyby of Jupiter in March 2006 and an encounter with Pluto around Christmas 2012. The spacecraft will be as small and as simple as possible to minimise the risk of technical problems. The craft will be dominated by a large radio antenna required to broadcast data back to Earth during and after the encounter with Pluto. Since the Sun's energy is so weak at these enormous distances, the use of solar panels to generate electricity is not practical. Instead, power will be supplied by radioisotope

187

thermoelectric generators similar to these used on other deep space missions. The Pluto–Kuiper Express will only carry a few scientific instruments, probably comprising a visible camera, an infrared mapping spectrograph and an ultraviolet spectrograph. These should provide images showing details as small as about 1 km across as the spacecraft hurtles through the Pluto–Charon system. Assuming all has gone well, the spacecraft may then be directed towards one or more suitable Kuiper Belt objects. No special equipment will be carried for this phase of the mission, the scientific instruments designed for observations of Pluto should be well suited for exploring any other objects which the spacecraft might encounter.

At present, no specific trans-Neptunian object has been identified as a possible target. Although Dave Jewitt has conducted a search for objects close to the expected flightpath of the probe, the region of sky is close to the Milky Way where the star density makes searching difficult. He suggests that what is needed is a dedicated search programme, perhaps using NASA's proposed New Planetary Telescope, specifically in support of the Pluto–Kuiper Express mission. In any event, there is no rush to pick a target as another spanner was soon thrown into the works. In September 2000 NASA announced that it planned to expand further its exploration of the planet Mars and soon afterwards ordered that work on the Pluto–Kuiper Express be stopped. Although insisting that the stop-work order was only a postponement, and not a cancellation, astronomers interested in Pluto fear that the two things are effectively the same. Even a year's delay in the launch of the Pluto spacecraft will mean that it will no longer be able to use Jupiter's gravity to boost itself towards the outer solar system. Without such a boost the mission will take longer, or the already small spacecraft will have to be made smaller. Either way, its arrival at Pluto will be delayed by years, perhaps to the year 2020. This would be fatal to many of the mission's scientific objectives. Pluto is receding from the Sun and as it does so it is getting colder and colder. By 2020, perhaps sooner, the planet's tenuous atmosphere will have frozen onto the surface. Once this happens the atmosphere will remain trapped as ice for over 200 years. For studies of Pluto's atmosphere, time is rapidly running out.

Inevitably there was an outcry, and attempts to reverse the decision began. Astronomers argued with NASA while astronomy societies and individuals lobbied the US congress to have the project reinstated. However, they may have been too late; NASA's 2001 budget was approved in November 2000. Attempts to reverse the postponement will doubtless continue, and hopefully succeed, but only once the Pluto–Kuiper express is safely on its way to the launch pad will searches for its second stop become more important.

Will we ever get our names right?

CHAPTER 12

Amongst the many sincere and hotly debated issues concerning the origin, content and composition of the trans-Neptunian region are a few questions of little scientific importance. None the less these are issues of considerable interest. The first is no less fundamental than the name of this recently discovered region at the edge of the solar system; should it really be called the Kuiper Belt, and what should we call the objects within it?

The term Kuiper Belt first surfaced in the seminal 1988 paper by Duncan, Quinn and Tremaine in which they reported their conclusions concerning the probable source region of the short-period comets. They referred to this structure, which at the time was a purely theoretical concept, as the 'Kuiper Belt'. The name appears to have been coined by Scott Tremaine, although he says it arose more-or-less spontaneously since both 'Kuiper' and 'comet belt' appeared in the opening sentence of Julio Fernandez's 1980 paper on the origin of the short-period comets. Alan Stern had used the term in 1990 while describing his arguments for an ancient population of ice dwarfs, but the name attracted little attention until observational astronomers began to discover real objects in this region. Dave Jewitt and Jane Luu entitled the paper describing their first find as 'Discovery of the candidate Kuiper Belt object 1992 QB_1'. In this paper they spoke quite clearly of the new object as being the first known member of the Kuiper Belt. The name quickly caught on and before very long all the trans-Neptunian objects were being referred to as KBOs or Kuiper Belt objects. However, it was not long before somebody called foul!

Jack Lissauer, from the State University of New York, had mentioned Edgeworth's 1949 paper when he reviewed the process of planet formation in 1993 and Paul Weissman, a cometary scientist from the Jet Propulsion Laboratory, had made reference to Edgeworth's work in a review written in 1995. However, the most vocal arguments that more credit should be given to Kenneth Edgeworth came from astronomers in Britain. Prominent amongst these were a small number of people in Northern Ireland who were already familiar with Edgeworth's 1943 and 1949 papers. They, and others, pointed out that Edgeworth had predicted the existence of a trans-Neptunian disc and had even remarked that its denizens might enter the inner solar system to be seen as comets. This, they argued, was as good a prediction as one was likely to get. What was more, Edgeworth's papers came out before Kuiper had published anything on the subject. Kuiper's remarks appeared in 1951, in a chapter which he had contributed to a book edited by J. Allen Hynek.[†] Surely then, the bodies in the trans-Neptunian region should be in the Edgeworth Belt, not the Kuiper Belt? What is more, they noted, Edgeworth spoke of a disc of material, rather than the ring of material described by Kuiper. Fuelling the controversy was the fact that Kuiper's famous book chapter made no reference to Edgeworth's papers, ignoring the scientific niceties of citing the work of other workers in the same field. Why should this be?

Since both Gerard Kuiper and Kenneth Edgeworth are now dead, (Edgeworth died in Dublin on 1 October 1972, Kuiper on Christmas Eve 1973 in Mexico City) we may never know for certain. None the less, it is interesting to speculate on the circumstances which led to one of these men being virtually ignored, while the other seems likely to be immortalised as one of the few people to have not simply a crater, or a small celestial body, but an entire region of space carrying their name for the foreseeable future.

Firstly, there is the vexed question of why Kuiper did not refer to Edgeworth's papers in his 1951 review. Could it be that he didn't know about them? After all, Edgeworth's first paper was published, in greatly abbreviated form, in the middle of the worst war the world has ever seen. 'Not so', says Mark Bailey, director of the Armagh Observatory in Northern Ireland. Bailey points out that today, with

---

[†] Hynek was an accomplished astronomer, although is now probably best known for his categorisation of UFO sightings, one of which was, 'A Close Encounter of the Third Kind'.

many more scientists active and a plethora of scientific journals and electronic media in which to publish information, just keeping up with the flood of scientific literature is a major task in itself. This was hardly the case fifty or sixty years ago. In the 1940s and 1950s there were far fewer astronomical journals being published and those which were contained very few papers on solar system research, then considered an astronomical backwater of little interest to astrophysicists as a whole. Kuiper was undoubtedly a world leader in solar system astronomy and it would not have been difficult for him to keep abreast of the small number of papers being published in his field. Even if Edgeworth's 1943 note in the *Journal of the British Astronomical Association* had somehow slipped past him, could he have failed to notice the much larger paper in the 1949 *Monthly Notices of the Royal Astronomical Society*? Dan Green, writing in the *International Comet Quarterly*, remarks that in the 1950s the *Monthly Notices* was in the top three or four of the world's most-read astronomical journals and that it would be odd to think that Kuiper would not be aware of what was being published there. Dave Jewitt says much the same. 'Do you think Kuiper wasn't reading *Monthly Notices*?' he asks rhetorically, 'He must have been'. So did Kuiper ignore this paper deliberately and somehow steal Edgeworth's idea? To be fair, Kuiper never claimed the credit for his prediction. For one thing, he had been dead almost two decades before the first trans-Neptunian object was discovered and anyway, his reputation as a great solar system astronomer was already secure.

Gerrit Pieter Kuiper was born in the Netherlands on the 7th of December 1905. As a young man, he was an outstanding student whose astronomical interests were encouraged by his father and grandfather. It was they who gave him a small telescope with which to pursue his hobby. Kuiper was awarded a degree from the University of Leiden in 1927 and entered postgraduate studies immediately afterwards. His professors at Leiden included such famous astronomers as Ejnar Hertzsprung, Willem de Sitter, Jan Woltjer and Jan Oort. Kuiper's PhD thesis concerned binary stars and in 1933, his thesis complete, he moved to the Lick Observatory in California. After a brief stay at Lick he moved by way of Harvard University to the Yerkes Observatory, part of the University of Chicago. He continued to work in the area of stellar astronomy for many years.

In the winter of 1943–1944 Kuiper turned his attention briefly onto the planets. He found the first evidence that Saturn's satellite Titan has an atmosphere containing methane gas and this discovery diverted him to solar system research. Using new infrared detector technology developed during the Second World War he began a programme of infrared spectroscopy of the giant planets and their satellites. In 1948 he discovered Miranda, the fifth known satellite of Uranus and in 1949 he found Nereid, the second moon of Neptune. He left the Yerkes Observatory in 1960 and founded the Lunar and Planetary Laboratory in Tucson, Arizona, still a major research institute today. In the 1960s Kuiper worked on various NASA space projects, including the robotic Ranger and Surveyor lunar missions. He had few research students, claiming he was too busy to supervise them, but those who did study with him included some very accomplished planetary scientists of the next generation, including Tom Gehrels, William Hartmann, Carl Sagan, Dale Cruikshank and Toby Owen.

In parallel with his research projects, Kuiper was always interested in finding better sites for ground-based observations and was instrumental in establishing major observatories in both Chile and Hawaii. He was also in the forefront of developing infrared telescopes carried in high-flying aircraft. As a tribute to his work in so many areas of solar system astronomy, Kuiper's name is carried on one of the brightest craters on the planet Mercury, a large crater on Mars, on Minor Planet 1776 Kuiper and by NASA's Kuiper Airborne Observatory.

Although some people believe Kuiper had a bad habit of minimising the work of other people by making very selective use of references, scientists who knew him, including ex-students of his still active in planetary astronomy, offer other possible explanations for his failure to refer to Edgeworth's work. Firstly, they point out that the social structure of astronomy has changed in the last fifty years. Today, scientists often work in large teams, but then scientific papers with several authors were quite rare. Similarly, today most papers have comprehensive lists of references pointing to work done by other

Figure 12.1 Gerard Kuiper (with his belt) stands outside the Yerkes Observatory in the 1950s. (Yerkes Observatory.)

195

people in the field,[†] but this was not always the case. Secondly, Kuiper's interests spread across the gamut of problems in solar system astronomy. While he may have seen Edgeworth's papers, he may have already come to much the same conclusions himself, or have subliminally absorbed them and then later thought of them as mostly his own. According to Dale Cruikshank,[‡] Kuiper was a strong personality whose native brilliance was accompanied by an uncompromising assurance of the importance of his own work. Cruikshank thinks it highly likely that Kuiper knew about Edgeworth's papers, but that he just ignored them and developed the same ideas himself. This is an opinion supported by astronomer Willem Luyten, who knew Kuiper at the Lick Observatory. Luyten, writing in 1979, said that he had no doubts about Kuiper's intelligence, but felt that Kuiper was sometimes able to put completely out of his mind that somebody else had had an idea first and would simply adopt it as his own. Edgeworth's case may not be the only example. According to Luyten, a Dutch astronomer named H. P. Berlage, whom he had known years before as a fellow student, came to him in 1946 with a voluminous paper describing a model for the formation of the solar system from a turbulent solar nebula. Although Luyten warned his old acquaintance that the paper was too long, Berlage had no time to modify his manuscript. It was submitted unchanged to the *Astrophysical Journal*, whose editorial offices were at the Yerkes Observatory where Kuiper was the director. As Luyten had feared, the paper was not accepted and was eventually returned unpublished. However, Kuiper's chapter in Hynek's 1951 book described a theory for the origin of the solar system which closely paralleled Berlage's ideas. It is by no means impossible that Kuiper had seen Berlage's manuscript while it was at Yerkes.

Kuiper also had a reputation of never quoting the work of another scientist unless he knew them personally and had a good grasp of their abilities. Edgeworth was an unknown quantity, an amateur theoretical astronomer working in a foreign country. Perhaps Kuiper somehow didn't feel it was appropriate, or relevant, to make reference to his work. There is also a question of timing. Edgeworth's paper was received by the Royal Astronomical Society in June 1949 and appeared

[†] Whether this is to give due credit to other people, or merely to minimise the risk of offending potential referees who must approve the paper for publication is not always clear.

[‡] To whom I am indebted for a copy of his excellent biographical sketch of Kuiper originally published in the *Biographical Memoirs of the National Academy of Science*.

in one of the last issues of that year's journal. However, the issue was not actually printed until March of 1950. So it might not have reached the Yerkes Observatory until the late spring or early summer of 1950. Kuiper later claimed that the bulk of his chapter in Hynek's 1951 book was done as early as 1949, which would mean that it was written about the same time as Edgeworth's paper. However, Kuiper's chapter does include references to, and discussion of, the 1950 papers of Jan Oort and Fred Whipple. There is also a citation of one of Kuiper's own 1951 papers so it seems that at least some late updating had been done before the book went to press.

Kuiper's review, being in many other ways so comprehensive, may have distracted other workers away from the earlier literature. Julio Fernandez, who turned his attention to the problem of the short-period comets in the late 1970s, admits he never saw any of Edgeworth's papers while he was doing his research. Of course, by then Edgeworth's work had been forgotten for almost twenty years and Fernandez can perhaps be forgiven for not searching so far back into the literature. However, Kuiper and Fernandez were not alone in passing over Edgeworth's writings. By 1961, Edgeworth had written an entire book, *The Earth, The Planets and the Stars: Their Birth and Evolution*, which was published by the MacMillian Company in New York, but the very next year Alastair Cameron wrote a major review of theories of the origin of the solar system without mentioning Edgeworth. Cameron did however give at least passing mention to Kuiper's remarks about a trans-Neptunian comet population. Cameron was a nuclear physicist who had been working on problems associated with how elements are built up inside stars and his main interest in Kuiper's work was to do with Kuiper's treatment of gas dynamics in the outer solar system. He was not particularly inter-ested in the orbital distribution of comets and at the time had never heard of Kenneth Edgeworth. Looking back across a gulf of forty years, Cameron thinks that even if he had known about Edgeworth's book, its title would have suggested to him that it was a popular work and he would probably not have pursued it.

What about the British astronomical establishment? Surely they could hardly have failed to notice a paper in their most prestigious astronomical journal? Yet here too, Edgeworth's work sank into obscurity. Again we can only speculate, but perhaps the professional astronomers of the time did not like the idea of someone muscling in on their territory. This might have been especially true of someone who lacked a university education or a position in a recognised astronomical research group. The strong bonds that today link amateur astronomers, especially those working in cometary observing, with their professional counterparts were almost non-existent sixty years ago. Perhaps Edgeworth's writings were seen by the academic world as little more than irrelevant scribblings produced by someone on the fringes of real astronomy.

Another possibility is that for Edgeworth, the timing of his papers was just a bit too early. Edgeworth had described the objects in his putative disc as clusters, echoing then prevalent theories that comets were loose aggregations of particles, flying gravel banks so to speak. His paper appeared just before the now popular icy conglomerate model gained acceptance. So, in 1949, there was no reason to believe in solid icy objects in the outer solar system. However, Kuiper, when he wrote his review a little later, had the advantage of knowing about the work of Fred Whipple and Jan Oort, which was revolutionising the understanding of comets. Even so, according to Dan Green, it seems that Kuiper thought of the trans-Neptunian region mostly as a place where comets had been formed early on before being ejected by Pluto's gravity into the Oort Cloud. Kuiper does not appear to have considered that a trans-Neptunian reservoir of comets might exist today.

Edgeworth certainly seems to have lost interest in the astronomical establishment at some point. Although he was a visitor to various observatories in Ireland, and lectured to the Irish Astronomical Society about the origin of the solar system on 9th April 1951, he did not publish much further work after 1949. His later works were limited to his 1961 book and a couple of letters in *The Observatory* magazine. Despite his obvious interest in the subject, Edgeworth made little reference to astronomy in his autobiography, *Jack of all Trades, the Story of my Life*, which was published privately in 1965. In just two paragraphs, he merely noted his long-standing interest in the subject and that he had published a few papers and a book concerning 'Certain problems in astronomy'. That these papers deserved at least some consideration was eventually noted by the naming of Minor Planet 3487 Edgeworth in May 1999.

Whatever the merits of a case for posthumous recognition of Edgeworth's work, it may already be a losing battle in which a few people are simply fighting a desperate retreat. Astronomy is full of examples of things that got named after the wrong person and Edgeworth was not the first, and is unlikely to be the last, to lose out in this way. A classical example of this mis-crediting is probably the so-called Bode's Law. This set out a numerical sequence which, for a time at least, seemed to explain the relative distances of the various planets from the Sun. The relationship was actually discovered by Johann Titius von Wittenberg; it was merely publicised by the far better known, and so more influential, Johann Bode. In this regard, one could also point to the naming of the Oort Cloud. Although Oort's name is correctly associated with the idea of a distant comet cloud, many of his conclusions were presaged by Ernst Opik in 1932 and fellow Dutchman Van Woerkom in 1948. Dave Jewitt admits he has now become more aware of the work of Edgeworth and thinks that he, not Kuiper, deserves the credit for the prediction. Martin Duncan concedes that he doesn't know what his group would have done about a name for the comet belt had they known about Edgeworth's papers in 1988. None the less, the term Kuiper Belt has already entered common usage. Experience suggests that it is unlikely to be formally revoked or fall rapidly from grace.

In an attempt to rectify the perceived injustice of this situation, astronomers have tried to find a solution that respects both parties, and some refer to the region as the Edgeworth–Kuiper Belt. Others say, in effect, that since like it or not we have a Kuiper Belt, why not fill it with Edgeworth–Kuiper Objects, or EKOs? Some will not even go this far. Brian Marsden, for one, counters that neither Kuiper nor Edgeworth really got it right. It was, he says, Fred Whipple who first wrote realistically and quantitatively about a population of comet-like objects in the trans-Neptunian region. According to Marsden, Whipple sketched a comet belt extending from about 35–50 AU which stopped rather abruptly. Whipple said 'These things are about a 100 km in diameter and have V magnitudes of about 22. We have no hope of detecting them at present'. Marsden argues that this is closer to what we are actually talking about today and says, 'Neither Edgeworth or Kuiper wrote about anything remotely approximating to what we are seeing, but Fred Whipple did'. Julio Fernandez also credits Whipple with helping to disseminate ideas about a comet belt. Fernandez describes a paper written by Whipple in 1972 as being very influential in his own thinking about the trans-Neptunian region.

To further confuse the issue, Dutch journalist George Beekman added a new angle to the debate in 1999. Writing in the October issue of the Dutch magazine *Zenit*, he reported that Armin Otto Leuschner was quoted in the 14th April 1930 issue of the *New York Times* as

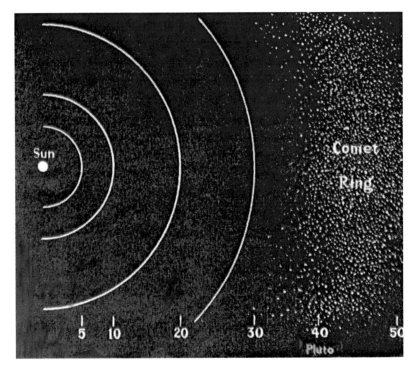

Figure 12.2 A sketch of a hypothetical comet belt published by Fred Whipple in 1964. Extending from 35–50 AU, Whipple's prediction was very close to what is now known as the Kuiper Belt. (F. Whipple.)

speaking of Pluto as being possibly a bright cometary object. Professor Leuschner was an astronomer at the University of California and he numbered amongst his students Fred Whipple and E. C. Bower. Based on the scanty information being released from the Lowell Observatory, Leuschner and his students had computed no less than six possible orbits for Pluto. In his remarks to the press, Leuschner seemed to have hedged his bets, remarking that Pluto might be an escaped main belt asteroid flung into the outer solar system by Jupiter, a long-period planetary object or a giant comet. Leuschner does not seem to have pursued the idea of a trans-Neptunian belt and, according to Marsden, Fred Whipple has no recollection of any discussion with Leuschner on the subject of a multiplicity of objects beyond Neptune. Marsden has however drawn attention to remarks by F.C. Leonard in an August 1930 leaflet published by the Astronomical Society of the Pacific. Leonard wrote, 'Now that a body of the evident dimensions and mass of Pluto has been revealed, is there any reason to suppose that there are not other, probably similarly constituted, members revolving around the Sun outside the orbit of Neptune?'. Leonard continued, 'Is it not likely that in Pluto there has come to light the first of a series of ultra-Neptunian bodies the remaining members of which still await discovery, but which are destined eventually to be detected?'. If this was a guess, it was a spectacularly good one.

Mark Bailey, one of Edgeworth's champions, raises another warning. He notes that since 1993 the expression 'Kuiper Belt' has come to cover a wide range of things including objects in resonances, those in the classical belt and those in the scattered disc. He cautions that, 'We should be careful about blurring important distinctions and try to use words with precision'.

Before moving on to more emotive, and even less scientifically productive issues, let us address one last question on the subject. Is it a belt, or a disc? Alan Stern much prefers to call it a disc. After all, he argues, the region is extended both radially outwards from the Sun and above and below the ecliptic plane. That is a disc, not a belt. Although Stern's argument is sound, it falls victim to the same problem of usage as before. The early papers of Duncan, Jewitt and others refer to the structure as a belt, and a belt it looks like staying. According to Hal Levison, Stern bowed to the inevitable when it was pointed out to him that if he continued to put Kuiper Disc, rather than Kuiper Belt, in the titles of his papers then computerised search engines were in danger of failing to find his work. If this were to happen, the citation rates of his papers would be reduced as no-one would ever read them. So, like it or not, it looks as if astronomy is stuck with a Kuiper Belt, albeit a thick one containing EKOs.

And what of these EKOs, what shall we call them individually? Traditionally, minor planet names have been suggested by the object's discoverer after observations spanning two or three oppositions have been made. By then the orbit is usually well-enough defined that the object is not likely to be lost again. The name is then approved by the Small Bodies Names Committee of the IAU. The Minor Planet Center uses the probable error in an object's orbit to determine when it might become ready to be named, but this scheme becomes more difficult when dealing with objects at great distances from the Sun. The slow motions and small observed arcs of these very-distant objects mean that it will be some time before they are going mathematically to reach the status of numberable minor planets (although the orbits for many of them are now quite secure) and when they do, it is quite possible that at least some of the original discovers might not even be alive to name them. There is also the question of their minor planet numbers. Should they be simply slotted into the regular list of minor planets? If this is done then there will be no way to distinguish a faint main belt asteroid from a faint Edgeworth–Kuiper object. A new system, using a new series of numbers prefixed with K was suggested, but this idea did not prove popular, and as of spring 2000, no Kuiper Belt object had been numbered, let alone named officially.

However, a few objects have been given unofficial names. None of these has yet come into general use, which in some cases is perhaps just as well. For a time, the Spacewatch team called the Centaur they discovered in 1992 'Big Red'. It is not a bad name, the object is quite big, and it is certainly red, but such a sobriquet was never going to catch on officially. The object was subsequently named Pholus.

Centaur 2060 Chiron has had its ups and downs as well. Although a few people had suggested quite early on that Chiron was an inert comet, the absence of any actual cometary activity around the time of its discovery meant that it was originally classified as an asteroid. The onset of cometary activity a few years later presented something of a problem. If Chiron was a comet, then perhaps it should have a comet's name and be known as Comet P/Kowal. This dilemma was solved by Mike A'Hearn, a respected cometary astronomer from the University of Maryland, who suggested that the object have some kind of dual status. As a result, 2060 Chiron is also designated 95P/Chiron, the 95P denoting that it is also the 95th comet proven to be periodic. How astronomers actually refer to the object often depends on what aspect of its behaviour they are investigating. Sometimes it is comet 95P/Chiron and other times just plain old Chiron.

The saga of unofficial names also extends to the first few trans-Neptunian objects. Soon after its discovery, news circulated that 1992 QB_1 was being called 'Smiley'. There was a grain of truth in this. Smiley was indeed the name Jewitt and Luu had given to the object when they found it. Luu was reading one of the spy novels of John Le Carre and, needing to call their discovery something for their own book-keeping purposes, they picked Smiley, Le Carre's spymaster. It seemed a reasonable name for something that had remained hidden for a long time and was only tracked down after much effort. Following the logic of this, they called the second object (1993 FW) Karla, codename of Smiley's main opponent. Although Luu says that the names were only intended for their own use as they tried to keep track of their various candidate objects, the more conservative elements in the astronomical community were not amused. It was very quickly pointed out that there was already an asteroid called Smiley. It is Minor Planet 1613, an otherwise obscure main belt object about 22 km in diameter which was discovered in 1950. It was named after Charles Hugh Smiley, a celestial mechanician who died in 1969. There is also a Minor Planet 1470 Carla and this similarity would more-or-less rule out calling something Karla. However, Smiley and Karla were never formally proposed to the IAU and when asked about it some time ago Dave Jewitt said, 'Let's not talk about names'. Jane Luu confirmed that in her opinion, 'There are better things to worry about than the names'.

However, Smiley and Karla would probably have been quite acceptable to the IAU compared with the informal names being used by Williams, Fitzsimmons and company when they discovered 1993 SB and 1993 SC. Talking around the subject at the observatory cafeteria, over what was probably one glass of wine too many, they reasoned that since Pluto was the name of a cartoon dog, the objects past Pluto might logically be named after cartoon cats. They soon abandoned the idea, but for a while the outer solar system was populated, unofficially at least, not just by Smiley and Karla, but by Felix and Garfield as well.

While the naming of individual objects is a matter for the discoverers and the appropriate bodies of the IAU, and the choice between Kuiper Belt, or Edgeworth Sheet, or even Duncan's Donut, will most probably be made by the custom and practice of astronomers active in the field, there is one naming issue which certainly caught the attention of the press and public. It concerned the planetary status of no less an object than Pluto.

The issue arose when it became clear that Pluto was just one of a number of objects in the 2:3 Neptune mean motion resonance. In many respects, Pluto's orbit is quite indistinguishable from those of other Plutinos and some astronomers soon began to refer to Pluto as simply the largest of the known objects in the Kuiper Belt. Brian Marsden had remarked on this as early as 1992 when discussing the description of 1992 QB_1 as the first Kuiper Belt object. He told the Boston Globe that, 'It was probably unfortunate that Pluto has been considered a planet', and asked, 'Is 1992 QB_1 the first Kuiper Belt object, or is Pluto the first?'. Things began to heat up when the increasing rate of discoveries from automated search telescopes meant that the number of numbered minor planets was rapidly heading towards 10 000. Since there was a tradition of naming aster-oids with 'round' numbers after someone or something special, such as 1000 Piazzi, 2000 Herschel and so on,[†] Brian Marsden suggested that Pluto might be numbered as minor planet 10 000. This would recognize Pluto's status as part of the Kuiper Belt. Marsden discussed the issue with members of the appropriate naming committee who agreed that this idea was viable. Mike A'Hearn, who had conceived the dual status idea for Chiron a few years earlier, was in favour of Pluto having dual status too. The idea of numbering Pluto soon began to circulate within the wider solar system community.

It was not a new issue. The question of Pluto's status as a real planet had come up from time to time before, especially when it was realised that the planet was much smaller than first thought. However, the issue had never really come to a head, perhaps because until the minor planet 10 000 idea surfaced, no-one had any better ideas. Marsden insists he never suggested demoting Pluto from the list of planets, suggesting merely that giving it a minor planet number would make things tidier when the other trans-Neptunians began to be numbered and catalogued. The dual status idea was seen by him as a compromise between the physical reality of Pluto as a large trans-Neptunian object and its traditional identity as a planet. Despite his good intentions, the proposal ignited a firestorm of criticism which, like most such debates, generated more heat then light.

On the one hand stood the group in favour of numbering Pluto. Some of their arguments ran as follows. Pluto is orbitally indistin-guishable from the other objects in the 2:3 resonance. Indeed, if you plot the objects in orbital element space, only an expert could pick Pluto out from the other objects in resonance with Neptune. Orbitally speaking, there is just nothing special about it. Although he would not go as far as saying Pluto should be demoted, as a dynamicist, Martin Duncan comes down firmly on the side of Pluto as a large Edgeworth–Kuiper object. Hal Levison agrees, saying, 'I firmly believe that if Pluto were discovered today we wouldn't be calling it a planet'. This sentiment is echoed by Brian Marsden, who rules that if discovered today Pluto would have got a temporary designation, then a minor planet number, just like any other similar object.

[†] The others are 3000 Leonardo, 4000 Hipparcus, 5000 IAU, 6000 United Nations, 7000 Curie and 8000 Isaac Newton. Minor Planet 9000 broke with tradition a bit, it was called Hal after the talking computer in the film *2001: A Space Odessy*.

Next comes the issue of Pluto's size. When it was first discovered Pluto was thought to be much bigger than we now know it to be. Indeed, the estimated size of Pluto has shrunk steadily for many decades and a light-hearted paper published in the 1980s showed that the decrease in the estimated size of Pluto could be fitted by a curve that would predict the planet's complete disappearance quite soon. On the question of size, Marsden says, 'If you want to consider a planet as something that is spherical and has collapsed under gravity, then you can have many such objects, fifteen, twenty, maybe more. There are plenty of main belt asteroids which would meet these standards.'. He describes the original naming of Pluto as a planet as 'irrational' and feels that it only came about because the Lowell observatory did a very good public relations job of saying that they had found the planet predicted by Percival Lowell. Of course, we now know that Pluto was not Lowell's Planet X. Pluto was found by chance, or rather by reason of the very careful search that Tombaugh had made.

On the other side of the divide stood the traditionalists who argued, often passionately, that Pluto's status as a planet should not be imperilled. It is clear that the Lowell observatory remains very proud of 'its' planet and Robert Millis, director of the observatory, says that numbering Pluto makes, 'No sense', remarking that 'Pluto deserves to be considered as more than a minor planet'. Alan Stern, a staunch defender of Pluto's planetary status, agrees. He says that, 'It's not as if the Lowell staff knew Pluto was only 1000 km or so in size when they found it and they were pushing it on people. They really thought they had discovered a new planet'. Whatever we know now, the planetary lobby say, Pluto has been classified as a planet for over 60 years and to reclassify it now would be foolish and a break with tradition. Jim Scotti of Spacewatch agrees, 'People categorise things to bring some kind of order to things and Pluto was classified as a planet so let's keep it that way'. Marsden counters that this argument is not strictly correct. What we now call Minor Planet 1 Ceres was once happily designated the eighth planet (Neptune had not yet been discovered) and the next few asteroids were regarded as planets too. It was only with the discovery of more and more asteroids that Ceres, Vesta, Pallas, Juno were reclassified as a minor planets. Indeed, astronomy textbooks published as late as 1847 referred to eleven primary planets.

Continuing the defence of Pluto's planetary status are those who point out that Pluto is different from the rest of the objects in the 2:3 resonance in several important ways. Firstly, it is much bigger; with a now well-established diameter of 2300 km and a mass 0.00237 times that of the Earth. Pluto is small in planetary terms, but still several times bigger than its nearest rival amongst the Plutinos. The completeness of the searches by Tombaugh and Kowal makes it very unlikely that a brighter Plutino will ever be found. Secondly, Pluto has a satellite, and no other trans-Neptunian object has one. This is no longer a clinching argument as some main belt asteroids such as 243 Ida are now known to have tiny satellites, but Pluto's satellite Charon is large. In fact, Charon is so large compared with Pluto that many astronomers refer to the system as the 'Pluto–Charon binary' regarding it as a sort of double planet rather than an ordinary planet plus a moon. Finally, Pluto has an atmosphere, albeit a thin one that will freeze out on the surface sometime in the early decades of the twenty-first century as Pluto recedes from the Sun. So there you have it say the traditionalists, Pluto is spherical, bigger than the other things around it, has a moon and an atmosphere. What more do you want to make it a planet? Alan Fitzsimmons is pro-planet, although he admits that Pluto is a bit of both. 'Schizophrenic' is the word he used.

The issue rumbled on for a while with Brian Marsden urging that Pluto get the coveted minor planet number 10 000, but overall sentiment was against him. As the debate spread, more and more astronomers pitched in. David Hughes from England said that, 'It was just astronomers admitting to what they have known for a long time', while others claimed demoting Pluto was 'stupid'. Soon the debate started to spread outside the tight-knit community of solar system astronomers via the press and internet, becoming simultaneously more heated and less informed as it did so. Eventually, the General Secretary of the IAU, Johannes Anderson, felt obliged to make a formal statement of the IAU's position. Although thought to be personally in favour of numbering Pluto, on 3rd February 1999 Anderson stated that, 'No proposal to change the status of Pluto as the ninth

Figure 12.3 Pluto and Charon. Kuiper Belt objects or double planet? (Gemini Observatory.)

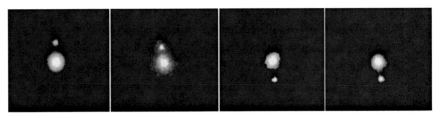

planet in the solar system has been made by any Division, Commission or Working Group of the IAU responsible for solar system science'.

Despite this announcement, various attempts were made to mount a vote on the issue. Many people felt this was not a sensible way forward. Alan Stern remarked, 'If we were to take a vote and rename Brian [Marsden] an amateur astronomer it wouldn't change the things he has done in his career, but it would be pejorative. It would taint things.' Marsden says he never meant to be pejorative. He thinks that Pluto is a most interesting object, but that it's just not a planet. None the less, votes were taken. One conducted by the Minor Planet Center came out strongly in favour of numbering Pluto, although an informal poll taken at a meeting of asteroid astronomers in Germany during 1998 came out with the opposite result.[†] According to Mike A'Hearn the debates have been remarkably emotional and the most interesting conclusion of the whole affair is the demonstration that astronomers are less rational than he thought they were.

Part, indeed most, of the problem is that there is no formal definition of a planet. Furthermore, it is very difficult to invent one which would allow the solar system to contain nine planets. Alan Stern feels that moving away from the rhetoric and actually defining what makes something a planet would help to crystallise our thinking. He suggests that for an object to be classified as a planet requires it to have three characteristics. It must be in orbit around a star (thus removing the larger satellites from contention), it must be too small to generate heat by nuclear fusion (so brown dwarf stars are excluded) and it must be large enough to have collapsed to a more-or-less spherical shape (which excludes comets, and most of the asteroids). These criteria would admit a few of the larger asteroids and probably some of the Kuiper Belt objects as well, but adding a requirement for a planet to have a minimum diameter of 1000 km would remove the larger asteroids from contention while retaining Pluto. However, setting a diameter of 1000 km is very arbitrary (why not use 1000 miles?) and it has no physical meaning in terms of how the objects formed or evolved. After all, if Pluto's companion Charon was just a bit larger, would it be called a satellite, or fully confirmed as half of a binary planet?

[†] The result from the German poll was 20 to 14 with a lot of abstentions.

Astronomers catalogue things to try and rationalise the Universe and to understand how it works. Objects which fall on some borderline, like the comet-cum-asteroid Chiron, can be studied by people with different perspectives and often provide crucial tests for cherished theories. From the point of view of the evolution of the solar system, it is sensible to consider Pluto along with the other trans-Neptunians, but from the point of view of how large solid bodies behave internally, Pluto is best considered a small planet. What the whole debate may be telling us is that there are at least three types of planets; rocky terrestrial planets like the Earth and Mars, giant planets like Jupiter and Neptune and a recently recognised class of ice dwarfs which encompasses Pluto, Charon, some of the large icy satellites and the large trans-Neptunian objects.

The arguments eventually subsided due to a mixture of scientific realism, tradition and perhaps an eye towards astronomy's public relations. From a scientific point of view it doesn't matter what we call Pluto. Reclassifying it won't actually help us understand its composition, origin or future. Jane Luu says, 'Sure, we can call it a planet, it's no skin off my back', as if to dismiss the issue as trivial compared with actually studying the object itself. Jewitt is firmly in the camp of those who regard Pluto as a large EKO, but says he is happy if people want to call it a planet. 'People are confused about lots of things', he says. Julio Fernandez is also in favour of the status quo. 'Astronomy is full of things with names that later proved to be incorrect or just plain silly', he points out. 'The lunar Maria, or seas, were named when astronomers thought they were just that, large bodies of open water.' Three hundred years later, when Neil Armstrong made his small step onto the moon he was wearing boots, not flippers, but no-one has suggested renaming the Maria. Probably no-one ever will. One other remark comes from Eileen Ryan who once shared an office with Clyde Tombaugh and who favours the status quo. 'Every time the subject came up Clyde was just crushed' she said.

Finally, there was the issue of the public perception of astronomy. Science progresses by continually investigating and challenging existing beliefs and sometimes finding out that earlier ideas were not quite as good as once thought. The idea that the Earth was the centre of the Universe held sway for centuries until a better idea came along. Even though it was correct to do so, demoting the Earth from its privileged position took some doing. Even in today's more rational times, for some people giving up an idea held dear for many years is difficult.

It seemed to some that there was a real risk that a media interested in scandal and controversy would distort the issue and imply that the scientists had somehow, 'Got it wrong again'. Although Mark Bailey described the issue as a great opportunity to educate people about science, and Dan Green said it was time to stop teaching our children an outdated 1940s picture of the solar system, at least a few astronomers feared demands for the withdrawal of textbooks and presumably the firing of the incompetents who could not tell what a planet looked like.

The witching hour came and went when minor planet number 10 000 was finally reached. It was assigned without fanfare to a small main belt asteroid which was named Myriostos, which means 10 000th in Greek. Of course astronomers may live to regret this decision if an object bigger than Pluto is ever found amongst the classical Kuiper Belt or, more likely, in the scattered disc or the Oort Cloud. Certainly there is no reason why a few large Kuiper Belt objects might not exist in the dark outer reaches of the solar system. The gravitational forces which eject objects into the disc do not care how big they are, as long as they are small compared with one of the giant planets. So a Pluto-sized, or even-Earth sized object may well be lurking out in deep space waiting to be discovered. If one ever is, then Brian Marsden may have the last laugh. Perhaps he even has a bet on it with someone.

Dramatis personae

Many of the astronomers mentioned in this story have been honoured by having a minor planet named after them. The citations for those minor planets are given here. They were true at the time of the naming, but in at least some cases they are no longer up to date. To preserve the flavour of the original citations, I have not attempted to update them. It is important to recognise that minor planet naming is not a systematic process and so this list is by no means comprehensive. There are certainly astronomers who have contributed mightily to this field and not yet been assured of a place in the heavens. Here are a few who have.

(3192) A'Hearn

1982 BY₁. Discovered 1982 January 30 by E. Bowell at Anderson Mesa

Named in honor of Michael F. A'Hearn, professor of astronomy at the University of Maryland. A prominent student of cometary physics, A'Hearn has pursued coordinated spectroscopic and spectrophotometric observations of comets spanning the spectral interval from the vacuum ultraviolet to the radio region. He participated in the 1983 discovery with the IUE spacecraft of diatomic sulphur in the spectrum of Comet IRAS–Araki–Alcock (1983d) and has made many other important contributions to our current understanding of comets.

(4050) Mebailey

1976 SF. Discovered 1976 September 20 by C.-I. Lagerkvist and H. Rickman at Kvistaberg

Named in honor of Mark E. Bailey, a British astronomer at the University of Manchester well known for his work on the origin of comets, the dynamics of the Oort Cloud and the capture of comets into short-period orbits.

(3485) Barucci

1983 NU. Discovered 1983 July 11 by E. Bowell at Anderson Mesa

Named in honor of M. Antonietta Barucci, planetary scientist at the Istituto di Astrofisica Spaziale in Rome. A prolific contributor to the study of the physical properties of minor planets, Barucci has carried out both photometric and astrometric observations at the telescope and has studied minor planet body shapes and surface light-scattering properties in the laboratory.

(7553) Buie

1981 FG. Discovered 1981 March 30 by E. Bowell at Anderson Mesa

Named in honor of Marc W. Buie (1958–), an astronomer at Lowell observatory, who has made many important contributions to planetary astronomy. These include the identification of water ice on Pluto's satellite Charon, constraints on the albedo and frost distribution on Pluto and Charon and improvements in our knowledge of Charon's orbit. Buie is also a codiscoverer of several trans-Neptunian objects and has developed a wide variety of astronomical software used at Lowell Observatory and elsewhere.

(3327) Campins

1985 PW. Discovered 1985 August 14 by E. Bowell at Anderson Mesa

Named in honor of Humberto Campins, research scientist at the Planetary Science Institute in Tucson. Well known for his work on the properties of cometary comae, Campins has helped establish pioneering techniques to measure the physical properties of cometary nuclei using simultaneous infrared and visual observations. He has also undertaken infrared searches for intramercurial bodies.

(4551) Cochran

1979 MC. Discovered 1979 June 28 by E. Bowell at Anderson Mesa

Named in honor of William D. Cochran and Anita L. Cochran, husband and wife astronomers at the University of Texas at Austin.

William's broad range of research has concerned planetary (including cometary) atmospheres, Raman scattering, stellar radial velocity variations and motions in stellar chromospheres and photospheres. Using a spectroscopic radial velocity meter, he is currently surveying several dozen stars to search for reflex motions (as small as about 2 m/s) that would indicate the presence of planetary companions. Anita is a specialist in the chemistry of cometary comae and in particular how the chemistry changes with changing heliocentric distance. An assiduous observer, she has used spatially resolved spectra to help transform the photometry of comets into a quantitative discipline. She has also developed sophisticated cometary models to understand how the observed atoms, molecules, and radicals are related to the larger parent molecules present in cometary nuclei. Anita is currently a team member of the Imaging Science Subsystem of the Comet Rendezvous – Asteroid Flyby mission.

(3531) Cruikshank

1981 FB. Discovered 1981 March 30 by E. Bowell at Anderson Mesa

Named in honor of Dale P. Cruikshank, planetary scientist at the University of Hawaii, Honolulu. Cruikshank is well known for his observational work on solar system small bodies, including Trojan asteroids, comets and Pluto. He is especially known for studies of outer-planet satellites, including Triton, Iapetus and Io, through both telescopic and Voyager spacecraft observations. He has been active in developing instrumentation and facilities at Mauna Kea Observatory and has promoted historical studies of planetary science. Through several extended working visits to the Soviet Union and other projects, Cruikshank has also been a leader in furthering international scientific relations.

(3638) Davis

1984 WX. Discovered 1984 November 20 by E. Bowell at Anderson Mesa

Named in honor of Donald R. Davis, senior scientist at the Planetary Science Institute in Tucson. Davis has made fundamental theoretical and experimental contributions to research on the collisional evolution of minor planets. With colleagues, he was the first to propose the 'gravitationally bound rubble pile' model for large minor planets.

Another of his research interests is infrared searching for intramercurial bodies.

(9064) Johndavies

1993 BH8. Discovered 1993 January 21 by the Spacewatch at Kitt Peak

John K. Davies (1955–) of the Joint Astronomy Centre was instrumental in the successful discovery and follow-up of (3200) Phaethon and several comets with the IRAS satellite in 1983. He has also carried out studies of the infrared nature of distant minor planets and authored a number of popular books and articles.

(6115) Martinduncan

1984 SR2. Discovered 1984 September 25 by B. A. Skiff at Anderson Mesa

Named in honor of Martin J. Duncan (1950–) of Queen's University, Kingston, Ontario. Duncan has made several important contributions to the understanding of the origin and dynamical evolution of small bodies in the solar system, particularly comets and the likelihood that they originated in the Kuiper Belt. He has been involved in the development of two important numerical algorithms that have led to orbital integrations of unprecedented duration.

(3487) Edgeworth

1978 UF. Discovered 1978 October 28 by H. L. Giclas at Anderson Mesa

Named in memory of Kenneth Essex Edgeworth (1880–1972), Irish engineer, economist, military man and independent theoretical astronomer, who reasoned that the solar system did not end with Neptune. As early as 1943 he pointed out the likely existence of a reservoir of potential comets near the invariable plane. This preceded the discovery of 1992 QB$_1$ by almost half a century.

(2664) Everhart

1934 RR. Discovered 1934 September 7 by K. Reinmuth at Heidelberg

Named in honor of Edgar Everhart (1920–1990), since 1969 in the physics–astronomy department at the University of Denver and director of the Chamberlain Observatory. After an impressive career working on atomic cross-sections, he has made equally fundamental contributions to our knowledge of the distribution of comets and the evolution of cometary orbits, including the development of an efficient integration technique for the purpose. Visual discoverer of comets 1964 IX and 1966 IV, he has more recently designed and constructed a measuring engine and used it in a highly successful program of photographic astrometry of comets.

(3248) Farinella

1982 FK. Discovered 1982 March 21 by E. Bowell at Anderson Mesa

Named in honor of Paolo Farinella, planetary scientist at the University of Pisa, whose research has included studies of the origin of the solar system and the dynamics of planetary satellites and ring systems. Farinella's work on minor planets has concerned the collisional evolution of the belt and the formation of families, both from a theoretical and an experimental point of view.

(5996) Julioangel

1983 NR. Discovered 1983 July 11 by E. Bowell at Anderson Mesa

Named in honor of Julio Angel Fernandez (1946–) of the Universidad de la Republica, Montevideo. Fernandez is a noted dynamicist who has worked on the evolution of comet orbits and planetesimal scattering in the outer solar system, including the formation of the Oort Cloud. His work has led to some of the first clear indications for the existence of the trans-Neptunian belt. Since 1985 he has contributed to the reestablishment of Uruguayan astronomy by educating a vigorous group of young planetary scientists and dynamicists.

(4985) Fitzsimmons

1979 QK$_f$. Discovered 1979 August 23 by C.-I. Lagerkvist at La Silla

Named after Alan Fitzsimmons, who works on the relationships between minor planets and comets and has collaborated with the discoverer for several years. His enthusiasm and good spirit when observing on La Palma has always been very much appreciated by the discoverer.

(1777) Gehrels

4007 P-L. Discovered 1960 September 24 by C. J. van Houten and I. van Houten-Groeneveld at Palomar

Named in honor of Tom Gehrels (1925–), staff member of the Lunar and Planetary Laboratory at Tucson. Dr Gehrels is well known for his photometric and polarimetric observations of minor planets and the Moon.

(7728) Giblin

1977 AW$_2$. Discovered 1977 January 12 by E. Bowell at Palomar

Named in honor of Ian Giblin (1969–), a British physicist who has performed a number of laboratory experiments to simulate hypervelocity impacts among minor planets. Giblin has developed new data analysis tools to study their outcome and to draw conclusions regarding the corresponding actual events.

(7638) Gladman

1984 UX. Discovered 1984 October 26 by E. Bowell at Anderson Mesa

Named in honor of Brett Gladman (1966–), a Canadian astronomer and dynamicist who has made important contributions to modelling the dynamical evolution of near-Earth objects and the transport of meteorites, including those from the Moon and Mars. Gladman has also carried out observational surveys of trans-Neptunian objects and in 1997 was codiscoverer of the two irregular satellites of Uranus.

(2068) Dangreen

1948 AD. Discovered 1948 January 8 by M. Laugier at Nice

Named in honor of Daniel W. E.Green (1958–), student aide at the Smithsonian Astrophysical Observatory during 1978 June–August, in appreciation of his invaluable assistance during the transition of the Minor Planet Center from Cincinnati to Cambridge [Massachussetts].

(3676) Hahn

1984 GA. Discovered 1984 April 3 by E. Bowell at Anderson Mesa

Named in honor of Gerhard Hahn, a planetary astronomer at Uppsala Observatory and a member of the research group studying minor planets and comets. Hahn has undertaken extensive photometry and astrometry of minor planets and has been studying the long-term orbital evolution and physical properties of these objects.

(3267) Glo

1981 AA. Discovered 1981 January 3 by E. Bowell at Anderson Mesa

Named in honor of Eleanor F. ('Glo') Helin, planetary scientist at the Jet Propulsion Laboratory, in appreciation of her extraordinary contributions to the discovery of near-Earth minor planets. Her finding of 1976 AA (=2062 Aten) heralded the recognition of a new class of planet-crossers, and her initiation of the Palomar planet-crossing asteroid survey has resulted in increased worldwide interest in the observation of minor planets. Helin's education and experience as a geologist and in the analysis of meteorites has provided a unique background for her interest in asteroids and comets.

(3099) Hergenrother

1940 GF. Discovered 1940 April 3 by Y. Väisälä at Turku

Named in honor of Carl William Hergenrother (1973–) of the Bigelow Sky Survey. This photographic survey has been very successful in discovering new high-inclination minor planets.

(4205) David Hughes

1985 YP. Discovered 1985 December 18 by E. Bowell at Anderson Mesa

Named in honor of David W. Hughes, reader in physics at Sheffield University, where he teaches courses on all aspects of astronomy. His research area concerns small solar-system bodies, particularly the relationship between comets and meteors. He has served astronomy in Britain as a vice president of both the Royal Astronomical Society and the British Astronomical Association. Hughes is a prolific reviewer of astronomy books and writes regularly on current issues in astronomy for *Nature*.

(6434) Jewitt

1981 OH. Discovered 1981 July 26 by E. Bowell at Anderson Mesa

Named in honor of David Jewitt (1958–) of the Institute for Astronomy, University of Hawaii. The consummate astronomer, Jewitt has been devoted to astronomy from a very early age. He has made several important contributions to planetary astronomy, starting with his discovery of the Jovian satellite Adrastea from Voyager data in 1979. He was co-recoverer of comet 1P/Halley in 1982. Jewitt is perhaps best known for co-discovering the first body in the Kuiper Belt in 1992 (see [minor] planet (1776)), thus proving that accretion occurs beyond the planetary region. Jewitt's main area of research is comets, but his wide-ranging interests have also produced work on planetary rings, minor planets, Pluto and circumstellar discs.

(1776) Kuiper

2520 P-L. Discovered 1960 September 24 by C. J. van Houten and I. van Houten-Groeneveld at Palomar

Named in honor of G. P. Kuiper (1905–1973), former Director of the Lunar and Planetary Laboratory at Tucson, and former Director of the Yerkes Observatory. Dr Kuiper was a well-known authority on the solar system and initiated both the McDonald Survey and the Palomar–Leiden Survey of minor planets.

(6909) Levison

1991 BY_2. Discovered 1991 January 19 by C. S. Shoemaker and E. M. Shoemaker at Palomar

Named in honor of Harold Levison (1959–) of the Boulder, Colorado, office of the Southwest Research Institute. Since 1988 Levison has, in collaboration with Martin Duncan, virtually revolutionised our view of the dynamics of short-period comets. The work uses sophisticated numerical models of test particles perturbed by the planets, and it has revealed important details about the Kuiper Belt, Centaurs, Pluto–Charon and short-period comets. Levison has also contributed revealing insights into the dynamics of perturbers in the Beta Pictoris system and was a leading member of the team that used Hubble Space Telescope to discover possible observational evidence for small comets in the trans-Neptunian region.

(5430) Luu

1988 JA$_1$. Discovered 1988 May 12 by C. S. Shoemaker and E. M. Shoemaker at Palomar

Named in honor of Jane X. Luu (1963–) for her research on the small bodies of the solar system. Luu is best known for her work with David Jewitt in discovering the first and subsequent members of the Kuiper Belt, as well as in following up with physical studies of those bodies. She has also contributed the most stringent upper limits on the existence of dusty comae around minor planets that might be dormant or extinct comets.

(6698) Malhotra

1987 SL$_1$. Discovered 1987 September 21 by E. Bowell at Anderson Mesa

Named in honor of Renu Malhotra (1961–), accomplished dynamicist and celestial mechanician at the Lunar and Planetary Institute in Houston. Born and raised in India, she has made major contributions to our understanding of how resonances affect satellite systems, the asteroid belt, and particularly Pluto. Malhotra was awarded the Harold C. Urey Prize by the Division for Planetary Sciences of the American Astronomical Society in 1997. Her talents and good spirits are much enjoyed by her colleagues.

(1877) Marsden

1971 FC. Discovered 1971 March 24 by C. J. van Houten and I. van Houten-Groeneveld at Palomar

Named in honor of Brian G. Marsden (1937–), Smithsonian Astrophysical Observatory, in recognition of his numerous contributions in the field of orbit calculations for comets and minor planets, his improved versions of the *Catalogue of Cometary Orbits*, and his activities in the Central Bureau and in Commission 20 of the IAU.

(4367) Meech

1981 EE$_{43}$. Discovered 1981 March 2 by S. J. Bus at Siding Spring

Named in honor of Karen J. Meech (1959–) of the Institute for Astronomy of the University of Hawaii for her pioneering studies of comets very far from the Sun. Her work following new and long-period comets to great distance has been a major factor in changing our ideas about water as the predominant driver of cometary activity in most comets. Meech's studies of (2060) = 95P/Chiron have likewise been critical in changing our understanding of the nature of the cometary coma.

(7639) Offutt

1985 DC$_1$. Discovered 1985 February 21 at the Oak Ridge Observatory at Harvard

Named in honor of Warren Offutt (1928–), on the occasion of his 70th birthday, 1998 February 13. After a career as an engineering executive, he turned in his retirement to the astronomical applications of CCDs, considering in particular the contributions that can be made by amateur astronomers. At his observatory in New Mexico he has made key observations of several of the objects discovered in the Kuiper Belt in recent years, as well as of other comets and minor planets as faint as 22nd magnitude. His follow-up of S/1997 U 2, one of the two recently discovered satellites of Uranus, played a crucial role in the establishment of its orbit.

(5040) Rabinowitz

1972 RF. Discovered 1972 September 15 by T. Gehrels at Palomar

Named in honor of the US astronomer David Rabinowitz and his work in the Spacewatch program.

(3594) Scotti

1983 CN. Discovered 1983 February 11 by E. Bowell at Anderson Mesa

Named in honor of James V. Scotti (1960–) of the University of Arizona, Tucson. Scotti works with the SPACEWATCH Telescope, which is the 0.9 m reflector of the Steward Observatory on Kitt Peak. He has developed most of the system's software and has carried out final checks and data reduction for the CCD scanning observations of comets and minor planets.

(4446) Carolyn

1985 TT. Discovered 1985 October 15 by E. Bowell at Anderson Mesa

Named in honor of Carolyn Spellmann Shoemaker, comet and aster-oid discoverer. Shoemaker began searching for asteroids in 1980, using plates taken at the UK Schmidt Telescope at Siding Spring. She helped develop a new photographic survey program using the 0.46 m Schmidt camera at Palomar Mountain and a newly designed stereomicroscope, which greatly increased the efficiency of film scanning. In 1983 Shoemaker found her first near-Earth asteroid, the Amor object (3199) Nefertiti, and later that year she found her first comet, 1983p. By February 1991 she had discovered 22 comets, at a rate of about one per 100 hours of scanning, and for discoveries recognised in the names of the comets she thus surpassed the tally of W. R. Brooks and moved into the all-time second place behind J.-L. Pons. Shoemaker already holds the record for finding new periodic comets: nine by early 1991.

(10234) Sixtygarden

1997 YB$_{8}$. Discovered 1997 December 27 by J. Tichß and M. Tichß at Klet

The street address of the Harvard-Smithsonian Center for Astrophysics is 60 Garden Street. Observers of minor planets and comets know it as the seat of the Minor Planet Center (see planet (4999)) and the Central Bureau for Astronomical Telegrams, which communicate fast-breaking news of astronomical discoveries to the international community.

(7554) Johnspencer

1981 GQ. Discovered 1981 April 5 by E. Bowell at Anderson Mesa

Named in honor of John R. Spencer (1957–), an astronomer at Lowell Observatory, for his pioneering interdisciplinary work in planetary science. Spencer's research includes detailed and insightful probing into the nature and character of the Galilean satellites. In particular, he has led the field in using high-resolution, ground-based imaging of Io to provide an excellent time history of volcanism, important for bridging the gaps between spacecraft encounters. In his studies, Spencer applies a keen intuitive sense of the natural world and leaves us with a better appreciation and understanding of our solar system.

(2309) Mr Spock

1971 QX$_{1}$. Discovered 1971 August 16 by J. Gibson at El Leoncito

Named for the ginger short-haired tabby cat (1967–) who selected the discoverer and his soon-to-be wife at a cat show in California and accompanied them to Connecticut, South Africa and Argentina. At El Leoncito he provided endless hours of amusement, brought home his trophies, dead or alive, and was a figure of interest to everyone who knew him. He was named after the character in the television program *Star Trek* who was also imperturbable, logical, intelligent and had pointed ears.

(6373) Stern

1986 EZ. Discovered 1986 March 5 by E. Bowell at Anderson Mesa

Named in honor of S. Alan Stern (1957–) of the Boulder, Colorado, office of Southwest Research Institute. Stern's research has focused on both observational and theoretical studies of the satellites of the outer planets, Pluto, comets, the Oort Cloud and the Kuiper Belt (see, respectively, planets (1691) and (1776)). He is also active in instrument development, with a strong concentration in ultraviolet and imaging technologies. He has participated in ten planetary sounding-rocket missions, two Space Shuttle mid-deck experiments and a Shuttle-deployable satellite. He was chair of NASA's Outer Planets Science Working Group during 1991–1994.

(4438) Sykes

1983 WR. Discovered 1983 November 29 by E. Bowell at Anderson Mesa

Named in honor of Mark V. Sykes, planetary scientist at the Steward Observatory of the University of Arizona, Tucson. Sykes was the first to suggest that the dust bands discovered in data from the Infrared Astronomical Satellite (IRAS) were due to the catastrophic disruptions of small asteroids and comets. He has also discovered several additional dust bands, a second type of dust trail, and identified parent comets responsible for some of the IRAS dust trails.

(3255) Tholen

1980 RA. Discovered 1980 September 2 by E. Bowell at Anderson Mesa

Named in honor of David J. Tholen (1955–), planetary scientist at the Institute for Astronomy of the University of Hawaii. Tholen's work on the eight-colour survey of minor planets led him to devise an improved taxonomy of minor planets. He has considered the physical properties of minor planets, satellites and comets in terms of composition and evolution, and he was among the first to observe events in the series of occultations and transits now occurring between Pluto and its satellite Charon.

(1604) Tombaugh

1931 FH. Discovered 1931 March 24 by C. W. Tombaugh at Flagstaff

Named by the Lowell Observatory after Clyde W. Tombaugh (1906–1997), the discoverer of Pluto, on the occasion of a symposium on Pluto, held on the fiftieth anniversary of its discovery, 1980 February 18. Tombaugh marked, during the course of his blink examination, over 4000 minor planets on plates obtained with the 0.33 m photographic telescope during the trans-Saturnian search program at the Lowell Observatory

(3634) Iwan

1980 FV. Discovered 1980 March 16 by C.-I. Lagerkvist at La Silla

Named in honor of Iwan P. Williams, of Queen Mary College, London, in recognition of his well-known work on meteor streams and interest in comets and minor planets. The discoverer appreciates their long and fruitful collaboration.

(1940) Whipple

1975 CA. Discovered 1975 February 2 at the Harvard College Observatory at Harvard

Named in honor of Fred L.Whipple (1906–), Harvard astronomer since 1931, professor since 1950 and director of the Smithsonian Astrophysical Observatory from 1955 to 1973. His countless contributions to our knowledge of the smaller bodies of the solar system include his icy-conglomerate model for cometary nuclei, and the development of modern techniques for the photographic observations of meteors. He has served as president of IAU Commissions 6, 15, and 22, and is now active on the NASA panel of space missions to comets and minor planets.

Index

Page numbers in italics refer to figures.

CPSIA information can be obtained at www.ICGtesting.com
Printed in the USA
BVOW071444081111

275605BV00004B/25/P

9 781107 402614